B. B. C. E. LIBRARY
950 Metropolitan Ave
Boston, MA 02136
(617) 364-3510

Ancillaries Available with
Practical English Handbook, NINTH EDITION

Print Ancillaries

- **Instructor's Annotated Edition.** The copy you are holding is the student edition of the text with answers to exercises overlaid in third color.

- **Practical English Workbook,** Fifth Edition, by Floyd C. Watkins, William B. Dillingham, and John T. Hiers. Collection of supplemental sentence and paragraph exercises that follows the organization of the handbook. Each unit opens with a brief grammar review. Accompanying **Instructor's Resource Manual** contains answers to exercises.

- **Diagnostic Tests.** Package of three forms of a 55-item diagnostic test on duplicating masters. Items and answers keyed to appropriate sections of the handbook.

- **Reference Chart.** Folded poster replicates table of contents and correction symbols of handbook for convenient grading.

Software Ancillaries

- **PEER: Practical English Exercises and Review.** Interactive program for practice in sentence editing. Items elicit five kinds of responses from students; explanations are cross-referenced to the handbook. Available in IBM and Macintosh.

- **Diagnostic Tests.** Computerized versions of the printed diagnostic tests. Available in IBM and Macintosh.

Offered at Special Prices

- **Practical English Handbook** and **The American Heritage Dictionary,** Second College Edition

- **Practical English Handbook** and **Fine Lines:** *Planning, Drafting, and Revising on the Computer,* by William H. Koon and Peter L. Royston. **Fine Lines** is an innovative software program of seven modules designed to help students in all stages of the writing process. The Drafting and Revising module contains an on-line handbook with MLA and APA documentation forms.

As part of Houghton Mifflin's ongoing commitment to the environment, this text has been printed on recycled paper.

Houghton Mifflin Company ▪ Boston

Dallas · Geneva, Illinois   Palo Alto   Princeton, New Jersey

Boston, MA 02
(617) 304-2540

xii   Contents

Research   331

Preface

In creating this Ninth Edition of the *Practical English Handbook,* our effort has been to keep it a true handbook—that is, a concise guide rather than a sprawling tome. We believe that a true handbook of English is above all handy—easy to use, easy to find one's way around in. Every page has been examined and reexamined with this thought in mind. We have resisted the temptation to elaborate beyond practicality and to illustrate beyond necessity.

The philosophical underpinnings of the *Practical English Handbook* continue to be traditional; that which generations of teachers and students found important and effective finds a place in this book as does a traditional tone and style of instruction in which good taste, respect, and dignity are, we hope, apparent. Tradition does not preclude change and improvement, however. Through the years and various editions, the *Practical English Handbook* has incorporated much of the best of current pedagogical thinking in regard to the teaching of writing and has remained attuned to the needs and preferences of instructors and students.

Like earlier editions, then, the Ninth Edition of the *Practical English Handbook* is conspicuous for its handiness and a down-to-earth, common-sense approach to good writing. It remains a true handbook in size and purpose. It is traditional but up-to-date. We have added further instruction in the methods of APA documentation and included a model paper that uses those methods. New model papers also appear in the sections Writing

Papers and Writing About Literature. We have revamped certain sections, like Logic and Accuracy, for greater conciseness and effectiveness. Throughout, we have changed and improved exercises and illustrative sentences as necessary for enhanced instructional value.

Practical English Handbook, Ninth Edition, is accompanied by the following ancillaries:

- *Practical English Workbook,* Fifth Edition—A collection of exercises that follows the organization of *Practical English Handbook* but can be used independently. Instruction is grouped at the beginning of each part so that exercises are freestanding on perforated pages. An Instructor's Manual contains answers to the exercises.

- Diagnostic Test Package—Three forms of a fifteen-page test provided on photocopying masters, free to adopters.

- Computerized Diagnostic Test Package—Interactive software using the same items as in the three printed diagnostic tests for IBM and Macintosh computers. Available at no cost to adopters.

- Reference Chart—A complimentary poster based on the table of contents and correction symbols in *Practical English Handbook* for use in grading papers.

- PEER: Practical English Exercises and Review—Computerized exercises for IBM and Macintosh computers complement *Practical English Handbook* and give practice in the areas of grammar, sentence structure, punctuation, mechanics, and diction and style. Students locate errors in sample sentences and then enter corrections. Correct answers appear superimposed above the original sentence on the screen. The relevant formal precept from *Practical English Handbook* is cited with carefully tailored instructional commentary applying the rule to the specific sentence in the exercise. PEER thus helps to reinforce students' understanding of important grammatical principles. PEER is available at no cost to adopters.

Also available at modest cost is *Fine Lines: Planning, Drafting, and Revising on the Computer* by William H. Koon and Peter L. Royston. This innovative new software program offers assistance at all stages of the writing process through seven flexible modules: Freewriting, Topic Narrowing, Idea Processing, List Processing, Outlining, Drafting and Revising, and The Journals. It also contains MLA and APA documentation models. For more information, contact the Houghton Mifflin Company sales office serving your area.

With each new edition, we grow ever more keenly aware of how dependent we are on the good will, good minds, and good teaching of others, so many, in fact, that we cannot name them all here. To all those who have written to us scores of valuable suggestions and words of encouragement, we express our deepest gratitude. We thank as well our colleagues and students in the English Department of Emory University for their help and the Reference Department of the Woodruff Library, especially Eric and Marie Nitschke. Below are listed some of the many writers and teachers who have helped us in deciding what should be included in the Ninth Edition and what should not be.

Leslie Angell, Bridgewater State College, MA

Linda Adams Barnes, Austin Peay State University, TN

Richard Battaglia, California State University, Northridge

Barbara Brumfield, Louisiana State University, Alexandria

Judy Callarman, Cisco Junior College, TX

Alfred Calvin, North Hennepin Community College, MN

Nancy Cox, Arkansas Tech University

Mary Flores, Lewis-Clark State College, ID

Craig Frischkorn, Jamestown Community College, NY

Judith Halden-Sullivan, CUNY/LaGuardia, NY
Larry Hawes, Aiken Technical College, SC
Barbara Holstein, Belmont Technical College, OH
Philip Korth, Michigan State University
Renu Juneja, Valparaiso University, IN
Philip Luther, Raymond Walters College, OH
Randall Marquardt, Boston University
William Peirce, Prince George's Community College, MD
John Peters, California State University, Northridge
Luke Reinsma, Seattle Pacific University
Margaret Strickland, Faulkner State Junior College, AL
John Taylor, South Dakota State University
Cedric Winslow, Iona College, NY

Floyd C. Watkins
William B. Dillingham

Practical
English
Handbook

Grammar

1

Anyone who can use a language, who can put words together to communicate ideas, knows something about grammar without necessarily being aware of it. Grammar is the methodology of language. It is also the formal study of that system and its laws or rules. For example, a person who says "They are" is using grammar. A person who says that *they* is a plural pronoun and takes the plural verb *are* is using terms of grammar to explain the proper use of the language. The technical study of grammar can be pursued almost for its own sake, but this book introduces a minimum amount of the grammatical system—all that is needed for you to communicate effectively in writing and speech.

The Parts of Speech

Knowing the part of speech of a word in its context may be the most basic aspect of grammar.

The eight parts of speech are **nouns, pronouns, verbs, adjectives, adverbs, conjunctions, prepositions,** and **interjections**. Each of these is explained and illustrated below.

NOTE: Much of the time, the function of a word within a sentence determines what part of speech that word is. For example, a word may be a **noun** in one sentence but an **adjective** in another.

noun
↓
She teaches in a *college*.

adjective
↓
She teaches several *college* courses.

Nouns

Nouns are words that name. They also have various forms that indicate **gender** (sex—masculine, feminine, neuter), **number** (singular, plural), and **case** (see **Glossary of Grammatical Terms**). Nouns are classified as follows:

(a) **Proper nouns** name particular people, places, or things *(Thomas Jefferson, Paris, Superdome)*.

Alice Walker was born in *Georgia*.

(b) **Common nouns** name one or more of a class or a group *(reader, politician, swimmers)*.

Few *authors* write anonymously.

(c) **Collective nouns** name a whole group though they are singular in form *(navy, team, pair)*.

The *crowd* watched eagerly.

(d) **Abstract nouns** name concepts, beliefs, or qualities *(courage, honor, enthusiasm)*.

Her *love* of *freedom* was as obvious as her *faithfulness*.

(e) **Concrete nouns** name tangible things perceived through the five senses *(rain, bookcase, heat)*.

The *snow* fell in the *forest*.

Pronouns

Most **pronouns** stand for a noun or take the place of a noun. Some pronouns (such as *something, none, anyone*) have general or broad references, and they do not directly take the place of a particular noun.

Pronouns fall into categories that classify both how they stand for nouns and how they function in a sentence. (Some of the words listed below have other uses; that is, they sometimes function as a part of speech other than a pronoun.)

(a) **Demonstrative pronouns** refer to specific objects or people (see **demonstrative adjectives**, p. 9). They can be singular *(this, that)* or plural *(these, those)*.

demonstrative pronoun
↓

Many varieties of apples are grown here. *These* are winesaps.

(b) **Indefinite pronouns** do not refer to a particular person or thing. Some of the most common are *some, any, each, everyone, everybody, anyone, anybody, one, neither* (see **8d**).

indefinite pronoun
↓

Everyone likes praise.

(c) **Intensive pronouns** end in *-self* (singular) or *-selves* (plural). An intensive pronoun is used to emphasize a word that precedes it in the sentence.

intensive pronoun
↓

Only the dentist *himself* likes the sound of his drill.

intensive pronoun
↓

I *myself* will carry the message.

(d) **Reflexive pronouns** end in *-self* or *-selves* and indicate that the subject acts upon itself.

reflexive pronouns ──────────┐
↓

I hurt *myself*.

You should remember to protect *yourself*.

(e) **Interrogative pronouns** are used in asking questions: *who, whom, whose, which. Who* and *whom* combine with *-ever: whoever, whomever* (see **9i**).

interrogative pronoun
↓
Who was chosen?

(f) **Personal pronouns** usually refer to a person, sometimes to a thing (see **9h**). They have many forms, which depend on their grammatical function.

	SINGULAR	PLURAL
First person	I, me, mine	we, us, ours
Second person	you, yours	you, yours
Third person	he, she, it	they, them, theirs
	his, hers, its	

(g) **Relative pronouns** are used to introduce dependent adjective or noun clauses: *who, whoever, whom, whomever, that, what, which, whose* (see **9i** and **7j**).

relative pronoun
↓
Special rooms are provided for guests *who* need quiet rest.

Relative pronouns may function as connectives as well as stand for a noun.

relative pronoun—connective
↓
The director had not read the play *that* was chosen.

■ Exercise 1

Identify and distinguish between the nouns and pronouns in the following sentences.

      N   **P**   **N**

1. A person who completes college is expected to be educated;
      P **P**   **N**
actually, one who completes college only has been exposed
    N
to knowledge.

    P   **N**   **N**

2. No one was surprised when the wind blew branches from the
        N   **N**   **N**
dead maple tree onto the roof of the metal tool shed, denting
P
it badly.

     N   **N** **P**

3. The shoebill is an awkward-looking bird that is native to
         N
eastern Africa.

     N   **P**   **P**   **N**

4. A nutrient is something that nourishes, such as an ingredient
   N
in food.

     N   **P**   **P**   **P**   **N**

5. The director himself told those who wanted to join the choir
               P **P**   **N**
to audition to show him they could not only sing on pitch
                     N
but also sight-read with confidence.

Verbs

Verbs can assert an action or express a condition (see 3).

action
↓
Tall sunflowers *swayed* gracefully.

condition
↓
The tall plants *were* sunflowers.

A main verb may have helpers, called **auxiliary verbs,** such as *are, have, will be, do, did.*

auxiliary verb   *main verb*
↘   ↙
The strong wind *will bend* the sunflowers.

auxiliary verbs   *main verb*
↙ ↘   ↓
Leif Erickson *may have preceded* Columbus to America.

Linking verbs are verbs like *appear, become, feel, look, seem, smell, sound, taste.* They express condition. The most common linking verbs are the many forms of the verb *to be (is, are, was, were, be, being, been).* (See p. 477.)

linking verb
↓
The woman with the plastic fruit on her hat *is* an actress.

linking verb
↓
In the presence of so many toys the child *appeared* joyful.

Verbs are either **transitive** or **intransitive** (see pp. 47, 481). (For **verb tenses,** see 4. For **verbals,** see pp. 22–23, 481.)

■ **Exercise 2**

Identify the verbs in each of the following sentences.

1. The aroma of fresh popcorn <u>wafted</u> through the building.

2. The flight attendant <u>walked</u> among the passengers and <u>looked</u> for the ring.

3. The symphony <u>was performed</u> by a distinguished orchestra, and the audience <u>was</u> enthusiastic.

4. Hercules <u>performed</u> the twelve labors that Hera <u>demanded</u>.

5. A chance encounter with a celebrity <u>can be</u> an exciting experience.

Adjectives

Adjectives modify a noun or a pronoun. They limit, qualify, or make more specific the meaning of another word. Generally, they describe. Most adjectives appear before the word they modify. (See **10**.)

> *adjective*
> ↓
> *Kind* words promote peace.

Predicate adjectives follow linking verbs and modify the noun or pronoun that is the subject of the sentence.

> *subject*   *predicate adjective*
> ↓⌒   ⌒↓
> A nap in the afternoon is *restful.*

The three **articles** *(a, an, the)* are classified as adjectives.

Some **possessive adjectives** have forms that are the same as some **possessive pronouns:** *her, his, its, our, their, your.* (*Her, our, their,* and *your* have endings with *-s* in the pronoun form.)

Demonstrative adjectives, which have exactly the same forms as demonstrative pronouns, are used before the nouns they modify.

this dog, *that* dog; *these* dogs, *those* dogs

Indefinite adjectives have the same form as indefinite pronouns: for example, *any, each, every, some.*

Adverbs

Adverbs (like adjectives) describe, qualify, or limit other elements in the sentence. They modify verbs (and verbals), adjectives, and other adverbs.

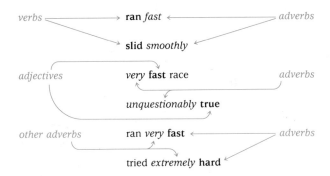

Sometimes adverbs modify an entire clause:

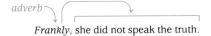

Frankly, she did not speak the truth.

Many adverbs end in *-ly (effectively, curiously),* but not all words that end in *-ly* are adverbs *(lovely, friendly).*

Adverbs often tell how *(slowly, well);* how much *(extremely, somewhat);* how often *(frequently, always);* when *(late, afterward);* or where *(there, here).* These are the main functions of adverbs, but not the only ones. Adverbs must be identified by their use in the sentence; that is, they cannot all be memorized.

■ Exercise 3

*Identify the adverbs and adjectives in the following sentences. Tell what each one modifies. (*A, an, *and* the *are adjectives.)*

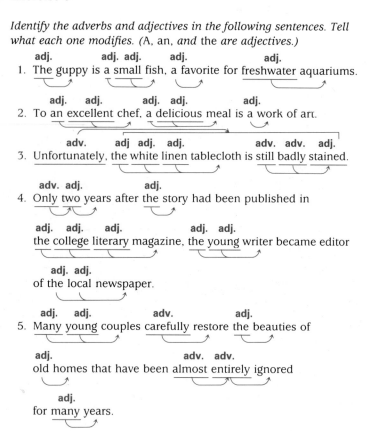

1. The guppy is a small fish, a favorite for freshwater aquariums.

2. To an excellent chef, a delicious meal is a work of art.

3. Unfortunately, the white linen tablecloth is still badly stained.

4. Only two years after the story had been published in the college literary magazine, the young writer became editor of the local newspaper.

5. Many young couples carefully restore the beauties of old homes that have been almost entirely ignored for many years.

Conjunctions

Conjunctions connect words, phrases, or clauses. They are classified as **coordinating, subordinating,** or **correlative.**

Coordinating conjunctions connect elements that are—grammatically speaking—of equal rank. Those most frequently used are *and, but, for, nor, or, so, yet.*

coordinating conjunction

The orchestra played selections from *Brahms, Bach,* **and** *Wagner.*

coordinating conjunction

The conductor left the stage, **and** *he did not return.*

Subordinating conjunctions introduce a subordinate or dependent element of a sentence. Examples are *after, although, as, as if, because, before, even though, if, in order that, once, since, so that, though, unless, until, when, where, while.*

subordinating conjunction

Although *many painters sell their work,* few become wealthy.

Some words (such as *before, after*) may function as conjunctions and also as other parts of speech.

Correlative conjunctions are always used in pairs. Examples are *both . . . and, either . . . or, not only . . . but also, neither . . . nor.* (See **18b.**)

correlative conjunctions

Not only *a well-balanced diet* **but also** *adequate sleep* is needed for good health.

■ Exercise 4

Identify all conjunctions in the following sentences, and tell whether each is coordinating or subordinating or correlative.

1. Despite heavy rains, the tour guide insisted on visiting all the
   coordinating   **subordinating**
   sites, and he said no money could be refunded because

   the schedule had been met.
   subordinating
2. When the sun comes up, the world takes on an entirely new
                          coordinating
   appearance, and hope seems to bloom again.
                                              coordinating
3. The ceilings of the old room were not high, but the windows
                   coordinating
   were wide, and fresh scents of flowers streamed through from

   the garden.
   subordinating
4. Although the telephone rang repeatedly, the desk clerk con-
                          coordinating   **coordinating**
   tinued to read a newspaper and occasionally to grunt or snort.
                                              correlative
5. The mayor recommends an increase not only in property
              correlative   **subordinating**
   taxes but also in wage taxes unless the budget deficit

   decreases dramatically.

Prepositions

Prepositions connect a noun or a pronoun (the **object of the preposition**) to another word in the sentence.

Most prepositions are short single words: *above, across, after, against, along, among, at, before, behind, below, beneath, beside, between, by, from, in, into, of, on, over, through, up, upon, with, without.*

Groups of words can also serve as prepositions: *along with, according to, in spite of.*

A preposition introduces and is part of a group of words, a phrase, that includes an object. The phrase is used as a unit in the sentence; a prepositional phrase acts as a single part of speech—usually an adjective or an adverb.

above the second floor	*across* the tracks	*after* the game
by the telephone	*through* that door	*under* one roof

(See **37f** for a discussion of the idiomatic use of prepositions.)

■ Exercise 5

Identify prepositions, objects of prepositions, and prepositional phrases in the following sentences.

1. The soprano delivered the aria with supreme effectiveness, a combination of sweetness and sorrow that greatly moved the audience.

2. According to the rules of the game, the contestant at the front of the line must turn quickly and race toward the rear.

3. There on the shelf was the book the hostess had borrowed from her friend.

      P   **O**   **P**   **O**

4. Never in the history of the province had so fortunate an event

occurred.

    P   **O**   **P**   **O**   **P**

5. After a fast stroll along the deck, the passenger fell into a

    O   **P**   **O**

chair and slept for an hour.

Interjections

Interjections are words that exclaim; they express surprise or strong emotion. They may stand alone or serve as part of a sentence.

> *Ouch!*
> *Well,* that is another story.

Because of their nature, interjections are used more often in speech than in writing, which is generally more deliberate than spontaneous.

■ **Exercise 6**

Name the part of speech of each word underlined and numbered in the following sentences.

     1   *2*   *3*   *4*   *5*   *6*   *7* *8* *9*

Oh, people are not always kind, but that is not justification

for bitterness.

	10	11	12	13	13	
	Successful	athletes	and	recognizable	politicians	mingled

	14	15		16	17		18
	with	famous	writers	and prominent	actors	at the	opera,

19	20
which	opened last night.

1. interjection	5. adverb
2. noun	6. adjective
3. verb	7. conjunction
4. adverb	8. pronoun

9. verb	15. adjective
10. adjective	16. adjective
11. conjunction	17. noun
12. adjective	18. noun
13. noun	19. pronoun
14. preposition	20. verb

The Parts of Sentences

A sentence is a group of words that expresses a complete thought. The essential parts of a sentence are a **subject** and a **predicate**.

A **subject** does something, has something done to it, or is identified or described.

subject	*subject*	*subject*
↓	↓	↓
Birds sing.	*Songs* are sung.	*Birds* are beautiful.

A **predicate** expresses what the subject does, what is done to it, or how it is identified or described.

predicate	*predicate*	*predicate*
↓	⌐↓⌐	⌐↓⌐
Birds *sing.*	Songs *are sung.*	Birds *are beautiful.*

Simple subjects, complete subjects, compound subjects

The essential element of a subject is called the **simple subject**. Usually it consists of a single word.

simple subject
↓
The large *balloon* burst.

The subject can be understood rather than actually stated. A director of a chorus might say "Sing," meaning "You sing." But here one spoken word would be a complete sentence.

understood subject   predicate
[*You*] Sing.

All the words that form a group and function together as the subject of a sentence are called the **complete subject**.

complete subject
The large balloon burst.

Subjects can be **compound**; that is, two or more subjects can be joined by a conjunction and function together.

compound subject
Students and *faculty* cheered.

Of course pronouns as well as nouns may make up compound subjects.

compound pronoun subject
He and *I* took a stroll.

■ **Exercise 7**

Identify the subjects and the complete subjects. Tell whether each sentence has a simple subject or a compound subject.

C 1. The piano teacher and her youngest pupil played a duet at the recital.

S 2. The oak tree has long been a symbol of strength and endurance.

S 3. The watchful otter turned and went back to her young.

C 4. The Brooklyn Bridge and the Golden Gate Bridge are both well known.

C 5. The wide river and the rapidly flowing stream converge near a village.

Simple predicates, complete predicates, compound predicates

The single verb (or the main verb and its auxiliary verbs) is the **simple predicate.**

simple predicate
↓
Balloons *soar.*

simple predicate
Balloons *are soaring.*

The simple predicate, its modifiers, and any complements (see pp. 19–20) form a group that is called the **complete predicate.**

complete predicate
Balloons *soared over the pasture.*

When two verbs in the predicate express actions or conditions of a subject, they are a **compound predicate**—just as two nouns may be a **compound subject**.

compound predicate

Balloons *soared and burst.*

compound, complete predicate

Balloons *soared over the pasture and then burst.*

(For errors in predication, see **15b**.)

■ Exercise 8

Identify the verbs (including auxiliaries) and the complete predicates. Tell whether each sentence has a simple predicate or a compound predicate.

S 1. The profit margin on the item was unusually slim.

S 2. All day long the caravan moved through long valleys and over steep hills.

S 3. Plan for a bright and meaningful future.

C 4. Playgoers arrived at the theater early and waited eagerly for the box office to open.

C 5. Children at the picnic played softball, swam in the pool, and rowed on the lake.

Complements

Complements, usually a part of the predicate, complete the meaning of the sentence. They are nouns, pronouns, or adjectives. They function as predicate adjectives or predicate nominatives (both sometimes called **subjective complements**) and direct or indirect objects.

Predicate adjectives

A **predicate adjective** follows a linking verb (see p. 7) and modifies the subject, not the verb.

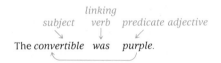

subject   *linking verb*   *predicate adjective*

The *convertible*   *was*   *purple*.

Predicate nominatives

A **predicate nominative** follows a linking verb and renames the subject. (Compare **predicate adjectives** above and **appositives, 23b.**)

predicate nominative

Winslow Homer was a *painter*.

Direct objects

A **direct object** receives the action indicated by a transitive verb. It is always in the objective case. (See **9.**)

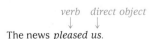

verb   *direct object*

The news *pleased us*.

Indirect objects

An **indirect object** receives the action of the verb indirectly. It is always in the objective case. The subject (through the verb) acts on the direct object, which in turn has an effect on the indirect object. Indirect objects tell to whom or for whom something is done. (See **9**.)

The sentence can be rearranged to read

When the preposition (*to,* as above, or *for*) is understood, the word is an indirect object. When the preposition is expressed, the word is an **object of a preposition:**

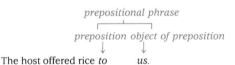

(Grammatically, the sentence above has no indirect object.)

Objective complements accompany direct objects. They can modify the object or rename it.

The editor considered the manuscript *publishable.*

The corporation named a former clerk its *president.*

■ Exercise 9

Underline and identify the predicate adjectives, predicate nominatives, direct objects, and indirect objects in the following sentences.

1. Those who possess **talents** (DO) should develop and use **them** (DO) wisely.

2. The gardener already had many **cactuses** (DO), but he was **happy** (PA) when a neighbor gave **him** (IO) some **more** (DO).

3. Good friends are **sympathetic** (PA) and **helpful** (PA).

4. Hand **applicants** (IO) a **brochure** (DO) and a **schedule** (DO).

5. The president will be a welcome **guest** (PN).

Phrases

A **phrase** is a group of words that does not have both a subject and a predicate. Some important kinds of phrases are **verb phrases, prepositional phrases,** and **verbal phrases.**

Verb phrases

The main verb and its auxiliary verbs are called a **verb phrase:** *were sitting, shall be going, are broken, may be considered. Were, shall be, are, may be,* and verbs like them are often auxiliary verbs (sometimes called "helping" verbs).

verb phrase

The seal *had been eating* the fish.

Prepositional phrases

Prepositional phrases function as adjectives or adverbs.

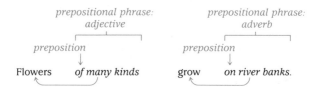

prepositional phrase:
adjective

preposition

Flowers *of many kinds*

prepositional phrase:
adverb

preposition

grow *on river banks.*

Verbals and verbal phrases

A **verbal** is a grammatical form derived from a verb. No verbal is a complete verb even though it may have an object and modifiers. (Like verbs, verbals can be modified by adverbs.) A verbal and the words associated with it compose a **verbal phrase.**

There are three kinds of verbals: **gerunds, participles,** and **infinitives.** A gerund is always a noun; a participle is an adjective; an infinitive may be a noun, an adjective, or an adverb, depending on its use in the sentence.

GERUNDS AND GERUND PHRASES

Gerunds always end in *-ing* and function as nouns.

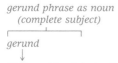

gerund phrase as noun
(complete subject)

gerund

Riding a bicycle is good exercise.

PARTICIPLES AND PARTICIPIAL PHRASES

Participles usually end in *-ing* or *-ed* (there are many irregular forms also) and always function as adjectives.

participle ending in -ing *(modifies* she)

Swimming steadily, she reached the shore.

participial phrase as adjective

participle ending in -ed *(modifies* bread)

Overbaked, the bread was hard.

INFINITIVES AND INFINITIVE PHRASES

Infinitives begin with *to,* which is sometimes understood rather than actually stated. They can be used as nouns, adjectives, or adverbs.

USED AS NOUN

infinitive phrase
used as subject

infinitive

To operate the machine was simple.

USED AS ADJECTIVE

infinitive phrase
modifies book *(noun)*

infinitive

Charlotte's Web is a good book *to read to a child.*

USED AS ADVERB

infinitive phrase
modifies reluctant *(adjective)*

infinitive

A true hero is reluctant *to boast of accomplishments.*

TO UNDERSTOOD

Someone must go [*to*] *buy* a tire.

■ Exercise 10

Identify phrases in the following sentences and tell what kind each is—infinitive, gerund, participial, prepositional, or verb.

        gerund
1. <u>Taking the census</u> is not as simple as it was two thousand

   years ago.

        prepositional   **verb**
2. Candidates <u>for local office</u> <u>will be attending</u> the parent-teacher
                infinitive   **infinitive**
   association meeting <u>to make themselves known</u> and <u>to express</u>

   <u>their views.</u>

               participial
3. <u>Perceiving his opponent's strategy</u>, the world-famous chess
             infinitive   **infinitive** **prepositional**
   player managed <u>to win the game</u> and <u>to do so</u> <u>in record time.</u>

  participial   **prepositional**
4. <u>Speaking strangely</u>, the young man <u>from the audience</u> said
                 prepositional   **infinitive**
   that his will was too strong <u>for him</u> <u>to be hypnotized</u>, but
              verb
   nevertheless he <u>was soon doing</u> whatever the hypnotist

   commanded.

         participial
5. <u>Insisting that his vegetables were fresh</u>, the gardener refused
   infinitive
   <u>to lower the price.</u>

Clauses

A **clause** is a group of words containing a subject and a predicate.
Clauses are either **independent** or **dependent (subordinate)**.

Independent clauses

An **independent clause** can stand alone gramatically and form a complete sentence. Two or more independent clauses in one sentence can be joined by coordinating conjunctions, conjunctive adverbs, semicolons, or other grammatical devices or punctuation marks.

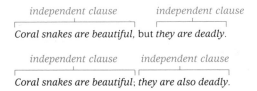

independent clause   *independent clause*

Coral snakes are beautiful, but *they are deadly*.

independent clause   *independent clause*

Coral snakes are beautiful; *they are also deadly*.

Dependent clauses

Like verbals, **dependent (subordinate) clauses** function as three different parts of speech in a sentence: nouns, adjectives, and adverbs. Unlike independent clauses, dependent clauses do not express a complete thought in themselves.

USED AS NOUN (usually subject or object)
That the little child could read rapidly was well known.
(noun clause used as subject)

The other students knew *that the little child could read rapidly*.
(noun clause used as direct object)

USED AS ADJECTIVE

Everyone *who completed the race* won a shirt.
(modifying pronoun subject Everyone)

USED AS ADVERB

When dusk comes the landscape seems to hold its breath.
(modifying verb seems)

The Kinds of Sentences

A **simple sentence** has only one independent clause (but no dependent clause). A simple sentence is not necessarily a short sentence; it may contain several phrases.

> Birds sing.
> The bird began to warble a sustained and beautiful song after a long silence.

A **compound sentence** has two or more independent clauses (but no dependent clause).

independent clause   *independent clause*

Birds sing, and *bees hum.*

A **complex sentence** has both an independent clause and one or more dependent clauses.

———*dependent clauses*———   ┌*independent*—

When spring comes and [when] new leaves grow, *migratory birds*

—*clause*—

return north.

A **compound-complex sentence** has at least two independent clauses and at least one dependent clause. A dependent clause can be part of an independent clause.

dependent adverb   *independent*
———*clause*———   ┌—*clause*—┐   ┌——*independent*——

When heavy rains come, *the streams rise, and farmers know*

dependent noun clause
used as object

———*clause*———

that there will be floods.

■ Exercise 11

Underline each clause. Tell whether it is dependent or independent. Explain the use of each dependent clause in the sentence. Tell whether each sentence is simple, compound, complex, or compound-complex.

complex   **independent**
1. In some towns people are awakened in the morning when a

                dependent adverb
factory whistle blows to signal the change in a shift of workers.

simple   **independent**
2. Moving down the hills and around the curves, the procession

of cars was not able to proceed faster than twenty miles an

hour.

simple   **entire sentence independent**
3. A samovar is not a Russian official but a large metal urn with

a spigot.

simple   **entire sentence independent**
4. On the Friday after Thanksgiving, in spite of heavy rain, we

took the train downtown to join the throngs of shoppers and

sightseers piling into the beautifully decorated department

stores.

compound   **independent**   **independent**
5. Events burst into the news, and then just as suddenly they are

replaced by other happenings.

Sentence Errors

1 Sentence Fragments *frag*

Write complete sentences.

Knowledge of what makes up a sentence (and what does not) is basic to an understanding of composition principles (see pp. 15–18). Writing a sentence fragment rather than a complete sentence can therefore be a serious error. Fragments take the form of dependent clauses, phrases, or other word groups without independent meaning and structure.

Notice how the fragments below are revised and made into complete sentences.

FRAGMENT (phrase)

Genealogy, the study of family history.

COMPLETE SENTENCE (verb added)

Genealogy is the study of family history.

FRAGMENT (dependent clause)

Although several large rivers have been cleaned up.

COMPLETE SENTENCE (subordinating conjunction *although* omitted)

Several large rivers have been cleaned up.

FRAGMENT (noun and phrase—no main verb)

The green fields humming with sounds of insects.

COMPLETE SENTENCE (modifier *humming* changed to verb *hummed*)

The green fields hummed with sounds of insects.

Fragments are often permissible in dialogue when the meaning is clear.

> "See the geese."
> "Where?" [fragment]
> "Flying north." [fragment]

Fragments are occasionally used for special effects or emphasis.

> The long journey down the river was especially pleasant. A time of
> rest and tranquillity.

■ Exercise 1

Make the following fragments complete sentences: delete one word;
add one word or a verb phrase; change one word; attach a fragment
to a sentence by changing punctuation and removing capitalization;
or rewrite.

1. The new educational standards ~~requiring~~ **require** a higher grade point

   average.

2. Medieval alchemists searched for the elixir of life **, w**/ ~~W~~hich was

   supposed to prolong life indefinitely.

3. Arbela was a city of ancient Persia **c**/ ~~C~~lose to where Alexander

   the Great defeated Darius III.

4. The Gallup poll is named for George Horace Gallup **, n**/ ~~N~~ot for

   a flagstaff in Gallup, New Mexico.

5. Lakes provide a wide range of recreation **, including w**/ ~~W~~ading, swimming,

   fishing, and boating.

6. Meadows of deep green grass, hills with trees soaring almost

   into the white clouds, one superhighway, and a small dirt

   road **make up** a landscape of contradictions.

7. ~~That~~ The bald eagle flies high over the mountaintops looking

   for prey.

8. An empty attic and an uncluttered basement suggest~~ive of~~ a

   life without a meaningful past.

9. The young artist ~~who~~ won first prize in the autumn exhibit

   of watercolors.

10. ~~If~~ **I** it becomes **may** possible to travel at a rate exceeding the speed

   of light.

2 Comma Splices and Fused Sentences   *cs/fus*

Join two independent clauses clearly and appropriately, or
write two separate sentences.

A **comma splice** or **comma fault** occurs when a comma is used
between two independent clauses without a coordinating conjunction.

SPLICE OR FAULT

    Human nature is seldom as simple as it appears, hasty judgments
    are often wrong.

A **fused sentence** or **run-on sentence** occurs when the independent clauses have neither punctuation nor coordinating conjunctions between them.

FUSED OR RUN-ON

> Human nature is seldom as simple as it appears hasty judgments are often wrong.

Comma splices and fused sentences fail to indicate the break between independent clauses. Revise in one of the following ways:

1. Use a *period* and write two sentences.

   > Human nature is seldom as simple as it appears. Hasty judgments are often wrong.

2. Use a *semicolon* (see also **22**).

   > Human nature is seldom as simple as it appears; hasty judgments are often wrong.

NOTE: Before *conjunctive adverbs* (see **20f** and **22a**), use a *semicolon* to join *independent clauses, or use a period and begin a new sentence.*

> Production of the item has greatly increased; *therefore,* the cost has come down.

> Production of the item has greatly increased. *Therefore,* the cost has come down.

3. Use a *comma* and a *coordinating conjunction (and, but, for, nor, or, so, yet)* (see also **20a**).

   > Human nature is seldom as simple as it appears, *so* hasty judgments are often wrong.

4. Use a *subordinating conjunction* (see pp. 11, 480) and a *dependent clause.*

> **Because** *human nature is seldom as simple as it appears,* hasty judgments are often wrong.

■ Exercise 2

Identify comma splices and fused sentences with CS *or* FUS. *Correct them. Write* C *by sentences that are already correct.*

CS   1. Russian music flowered late, however, it has developed rapidly

     during the past two centuries.

C   2. Few people realize that Sigmund Freud was born in what is

     now Czechoslovakia, where he spent his early years in poverty.

FUS   3. Conflicts almost always exist within a family, nevertheless it

     is still the most enduring of social units.

FUS   4. The bread stuck in the toaster, the smoke detector went off, the

     smell of burned toast permeated the apartment.

CS   5. Signs warning about riptides and the undertow were posted

     on the beach, no one ventured into the water.

■ Exercise 3

Identify each of the following as correct (C), *a comma splice* (CS), *a fused sentence* (FUS), *or a fragment* (F).

CS 1. The *Savannah* was launched in 1958, it was the first ship to be propelled by nuclear power.

C 2. Vitamins are necessary for health; however, excessive amounts of some of them are dangerous.

FUS 3. The magnitudes of earthquakes are measured by instruments called seismographs they record movements in the earth's crust.

F 4. Vitamins are necessary for health. Excessive amounts of some of them, however, dangerous.

C 5. Nearly every student of the classical guitar admires the playing of Andrés Segovia, who was essentially self-taught and who gave his first public performance when he was sixteen.

■ Exercise 4

Identify fragments (F), *comma splices* (CS), *or fused sentences* (FUS), *and correct them. Write* C *by correct sentences.*

CS 1. Continuing growth is necessary for a vigorous economy, reces-   *; or . R*

sions are therefore to be avoided if at all possible.

FUS 2. Jane Addams was a famous social worker she founded Hull-   *; or . S*

House in Chicago.

C 3. Dreams can be disturbing, but sometimes even unpleasant

dreams are to be preferred to reality.

FUS 4. Styles of clothing are not always planned for comfort belts   *; or . B*

and seams may be designed to look good rather than to fit

natural contours of the body.

F 5. Predictions were that videotapes of films would seriously

harm the motion picture industry. Another instance of how
— or , a

difficult it is to see into the future.

C 6. New York's Washington Square has changed considerably

since it was a haven for writers in the 1890s.

F 7. Snow ~~continuing~~ to fall, increasing the likelihood that classes
continued

would be canceled.

FUS 8. All over the city new buildings are springing up the skyline
; or . T

is rapidly changing.

C 9. For many, swimming is more pleasant than jogging, especially

when the weather is hot.

CS 10. An adventurer will sometimes participate in a pastime despite

its great danger, sky diving, for example, is perilous.
; or . S

■ Exercise 5

Follow the instructions for Exercise 4.

F 1. The tortoises on the Galápagos Islands often weigh great

amounts. Some as much as five hundred pounds.
, s

CS   2. After all, the student argued, any imbecile can punctuate *; or . S or , so* studying the mechanics of composition is a complete waste of time.

CS   3. Most generous people are naive *; or . T* they simply do not realize when they are being imposed upon.

FUS   4. Some families agreed that for two weeks they would keep their television sets turned off *; or . T* the children were surprisingly cooperative.

CS   5. Science and art are not incompatible *; or . S* some learned scientists are also philosophers or poets.

F   6. St. John's is the capital and port of Newfoundland *, a* an island off the eastern coast of Canada.

FUS   7. Finding someone to repair or clean mechanical watches is difficult *; or . M* most of the watches sold now are electronic.

C   8. For many people, microwave ovens have greatly shortened the time for preparing meals.

FUS   9. Why anyone would want to go over Niagara Falls in a barrel is puzzling *: or . M* many people have tried it, however.

F 10. Fossils and oil are often found in cold regions where ice and

                  This is e

    snow never entirely melt. ~~Ė~~vidence that the climate and the

    earth change drastically over long periods of time. *or* **rewrite:**
    Finding fossils and oil in cold regions where ice and snow never
    entirely melt proves that the climate and the earth change
    drastically over long periods of time.

3 Verb Forms *vf*

Use the correct form of the verb.

All verbs have three principal parts: the present infinitive, the past
tense, and the past participle. Verbs are regular or irregular.

    Regular verbs form the past tense and the past participle by
adding *-d* or *-ed* or sometimes *-t*. If only the infinitive form is
given in a dictionary, the verb is regular.

INFINITIVE	PAST TENSE	PAST PARTICIPLE
close	closed	closed
dwell	dwelled, dwelt	dwelled, dwelt
help	helped	helped
open	opened	opened
talk	talked	talked

Irregular verbs usually form the past tense and the past participle
by changing an internal vowel. For irregular verbs, a dictionary
gives the three principal parts and also the present participle. For
the verb *see,* the dictionary lists *see, saw, seen,* and *seeing.* For
think, it shows *think, thought* (for past and past participle), and
thinking (the present participle). The present and past participial
forms always have auxiliary (helping) verbs (see p. 7).

Know the following irregular verbs so well that you automatically use them correctly.

INFINITIVE	PAST TENSE	PAST PARTICIPLE
awake	awoke, awaked	awoke, awaked
be	was	been
begin	began	begun
bid (to offer as a price or to make a bid in playing cards)	bid	bid
bid (to command, order)	bade, bid	bidden, bid
blow	blew	blown
break	broke	broken
bring	brought	brought
build	built	built
burst	burst	burst
buy	bought	bought
choose	chose	chosen
come	came	come
deal	dealt	dealt
dig	dug	dug
dive	dived, dove	dived
do	did	done
drag	dragged	dragged
draw	drew	drawn
drink	drank	drunk
drive	drove	driven
drown	drowned	drowned
eat	ate	eaten
fly	flew	flown
freeze	froze	frozen
get	got	gotten, got
give	gave	given
go	went	gone
grow	grew	grown
know	knew	known
lead	led	led
lend	lent	lent
lose	lost	lost
ring	rang	rung
run	ran	run

INFINITIVE	PAST TENSE	PAST PARTICIPLE
see	saw	seen
sing	sang	sung
sink	sank, sunk	sunk
slay	slew	slain
sting	stung	stung
swim	swam	swum
swing	swung	swung
take	took	taken
teach	taught	taught
think	thought	thought
throw	threw	thrown
wear	wore	worn
write	wrote	written

Some verb forms are especially troublesome. *Lie* is confused with *lay*; *sit*, with *set*; and *rise*, with *raise*.

Lie, *sit*, and *rise* are intransitive (do not take objects) and have the vowel *i* in the infinitive form and the present tense.

Lay, *set*, and *raise* are transitive (take objects) and have *a*, *e*, or *ai* as vowels in the infinitive form and the present tense.

TRANSITIVE	lay (to place)	laid	laid
INTRANSITIVE	lie (to recline)	lay	lain
TRANSITIVE	set (to place)	set	set
INTRANSITIVE	sit (to be seated)	sat	sat

In special meanings the verb *set* is intransitive (a hen *sets*; the sun *sets*; and so forth).

TRANSITIVE	raise (to lift)	raised	raised
INTRANSITIVE	rise (to get up)	rose	risen

■ Exercise 6

Underline the incorrect verb and write the correct form above it.
Write C by correct sentences.

1. As the construction workers operated their jackhammers

   laid

   outside, the professor <u>lay</u> down her chalk, closed the window,

   threw

   and <u>throwed</u> her arms into the air.

   saw

2. She <u>seen</u> that she could not shut out the noise, so what she

   did

   <u>done</u> was to dismiss the class.

3. Although the most recently appointed member of the board

   came   **began**

   <u>come</u> in late to the meeting, she <u>begun</u> right away to make

   imaginative suggestions.

                                    took

4. Before the Bebo family bought a car, they <u>taken</u> few trips, but

                    used

   now that they are <u>use</u> to traveling, they seldom stay home in

   the summer.

           led

5. The builder <u>lead</u> the owner of the property to the back of the

                                        dragged

   lot and showed him where someone had <u>drug</u> old cars onto

   his land and had left them there as eyesores.

               gave

6. The passenger <u>give</u> a small tip to the cab driver, who frowned

   and glared.

         bid   **sitting**

7. One person <u>bidded</u> fifteen dollars, and then another <u>setting</u>

             bid

in the front row <u>bidded</u> twenty.

C 8. The artist laid the brush on the stand after he painted the

portrait, and it has been lying there ever since.

                         set

9. The carpenter promised to <u>sit</u> the bucket on the tile, but it

          sitting

has been <u>setting</u> on the carpet for a week.

               hung   **burst**

10. The balloon that <u>hanged</u> over the doorway suddenly <u>busted</u>.

4 Tense and Sequence of Tenses *t/shift*

Use appropriate forms of verbs and verbals to express time sequences. Avoid confusing shifts in tense.

The present tense expresses an action or condition occurring in present time; the past, an action or condition that occurred in past time but is now completed; the future, an action or condition expected to occur in future time.

4a Learn tense forms.

For each kind of time—present, past, and future—verbs have a different tense: simple, progressive, and perfect.[1]

---

    [1] There are in addition the emphatic forms with the auxiliary *do* or *did* (I *do go* there regularly).

	PRESENT	PAST	FUTURE
		Simple	
REGULAR	I walk	I walked	I shall (will) walk
IRREGULAR	I go	I went	I shall (will) go
		Progressive	
REGULAR	I am walking	I was walking	I shall (will) be walking
IRREGULAR	I am going	I was going	I shall (will) be going
		Perfect	
REGULAR	I have walked	I had walked	I shall (will) have walked
IRREGULAR	I have gone	I had gone	I shall (will) have gone

4b Be consistent in the use of tenses.

Relationships between verbs should be consistent.

TWO PAST ACTIONS

> The furniture maker *sanded* each piece with thorough care and only then *applied* [not *applies*] the stain.
>
> The child *smiled* broadly when she *saw* the clown approaching her.

TWO PRESENT ACTIONS

> As one *makes* new friends, school life *becomes* interesting.

4c Use the present tense for special purposes.

In addition to its usual function, the present tense is used for special purposes.

TO EXPRESS FUTURE ACTION

> The convention *begins* in four days. [simple present tense—future action]
>
> The convention *is beginning* in a few minutes. [present progressive tense—future action]

TO MAKE STATEMENTS ABOUT THE CONTENT OF LITERATURE
AND OTHER WORKS OF ART (historical present)

> In Henry James's *The Turn of the Screw,* a governess *believes* that
> the ghosts *are* real.

TO EXPRESS TIMELESS TRUTHS

> In 1851, Foucault proved that the earth *rotates* on its axis.

BUT

> Ancient Greeks *believed* that the earth *was* motionless.

Past tense is used for anything once believed but now disproved.

4d Use perfect tenses to indicate one time or action
completed before another.

The three perfect tenses are **present perfect, past perfect,** and
future perfect.

PRESENT PERFECT WITH PRESENT

> I *have paid* the rent; therefore, I **am moving** in.

The controlling time word sometimes is not a verb.

> I *have paid* the rent **already.**

PAST PERFECT WITH PAST

> I *had paid* the rent; therefore, I **moved** in.
>
> I *had bought* my ticket before the bus **came.**

FUTURE PERFECT WITH FUTURE

> I *shall have eaten* by the time we **go.** [The controlling time word, *go,*
> is present tense in form but future in meaning.]
>
> I *shall have eaten* by **one o'clock.** [Note that the controlling time
> words do not include a verb.]

The future perfect is rare. Usually the simple future tense is used with an adverb phrase or clause.

RARE

I shall have eaten before you go.

MORE COMMON

I shall eat before you go.

NOTE: The perfect participle also expresses an action that precedes another action.

Having laughed at length, the audience finally quieted down.

4e Use the present infinitive (*to* and the present tense of a verb) when it expresses action that occurs at the same time as that of the controlling verb.

NOT

I wanted *to have gone.*

BUT

I wanted *to go.*

NOT

I had expected *to have met* my friends at the game.

BUT

I had expected *to meet* my friends at the game.

NOT

I would have preferred *to have waited* until they came.

BUT

I would have preferred *to wait* until they came.

■ **Exercise 7**

Underline incorrect verbs or verbals, and write the corrections above them. Write C by correct sentences.

1. For years the foreign correspondent communicated with his
   took up
   parents by telephone, but then he <u>takes up</u> the habit of writing

   them long letters.

C 2. North American Indians are said to have originated the game

   of lacrosse.

3. In novels by Charles Dickens, a young man sometimes
   discovers
   <u>discovered</u> that his birthright is far more advantageous than

   he had imagined.

4. Having composed a rough draft on his computer, the author
   stared
   sat back in his chair and <u>stares</u> at the ceiling.
   headed
5. The horses rushed out of the starting gate and <u>head</u> for the

   first turn.

6. In looking back, public officials almost always say they would
   to remain
   have preferred <u>to have remained</u> private citizens.

7. The soprano met with the civic group, agreed to sing a solo
   began
   at the Christmas concert, and <u>begins</u> to search for suitable

   music.

8. The periodic table shows that the symbol for the element

   is

   mercury <u>was</u> Hg, not Me.

                                                              he began

9. Joseph Conrad was well into his thirties before <u>having begun</u>

   to write his novels.

                                                    has

10. In Gérôme's painting *The Cadet,* the young man <u>had</u> a slight

    sneer on his face.

5   Voice   *vo*

Use the active voice for conciseness and emphasis.

A transitive verb is either active or passive. (An intransitive verb does not have voice.) When the subject acts, the verb is active. In most sentences the actor is more important than the receiver.

PASSIVE   A positive impression *was made* on the employees by the new chief executive officer.

USE ACTIVE   The new chief executive officer *made* a positive impression on the employees.

PASSIVE   A brief but highly effective speech *was delivered* by the valedictorian.

USE ACTIVE   The valedictorian *delivered* a brief but highly effective speech.

In the sentences above, the active voice creates a more vigorous style.

When the subject is acted upon, the verb is passive. A passive verb is useful when the performer of an action is unknown or unimportant.

The book about herbs *was misplaced* among books about cosmetics.

The passive voice can also be effective when the emphasis is on the receiver, the verb, or even a modifier.

The police *were* totally *misled*.

■ Exercise 8

Rewrite the following sentences. Change from passive to active voice.

The white-handed gibbon quickly climbed the tree
1. ~~The tree was quickly climbed by the white-handed gibbon,~~

   and glared at the lions below.
   ~~and the lions below were glared at.~~

2. ~~Some ancient objects of art were discovered by~~ **T** the amateur

   discovered some ancient objects of art.
   archaeologist,∧

   reporter   **the road**
3. The ~~road~~ had ~~been~~ traveled ∧ many times ~~,by the reporter,~~ but

   she   **the old house.**
   ~~the old house~~ had never before ~~been~~ noticed ~~by her.~~

   The shipwrecked sailors witnessed a beautiful sunset,
4. ~~A beautiful sunrise was witnessed by the shipwrecked sailors,~~

   and at that very moment they spotted a ship dispatched
   ~~and at that very moment a ship dispatched to rescue them~~

   to rescue them.
   ~~was spotted by them.~~

The leaders of the tribe destroyed t   **used**
5. ∧The cameras that ~~were used by~~ the anthropologists∧ to take

pictures of the ancient village.~~were destroyed by the leaders~~

~~of the tribe.~~

■ Exercise 9

Change the voice of the verb when it is ineffective. Rewrite the sentence if necessary. Write E *by sentences in which the verb is effective.*

Mowing the grass and pruning the hedges
1. ~~The appearance of the yard was~~ dramatically improved ~~by~~
the appearance of the yard.
~~mowing the grass and pruning the hedges.~~

E 2. Both passengers were thrown clear, and they walked away

uninjured.

E 3. The horse lost the race because the shoe was improperly

nailed to its hoof.

Elderly people can have a   **they maintain**
4. ∧A good time ~~can be had by elderly people~~ if∧ relatively good

health.~~is maintained by them.~~

Cooking shows on television make s
5. ∧Some people ~~are made~~ hungry.~~by cooking shows on television.~~

6 Subjunctive Mood   *mo*

Use the subjunctive mood to express wishes, orders, and
conditions contrary to fact (see **Mood,** pp. 477–478).

WISHES

  I wish that tomorrow *were* here.

ORDERS

  The instructions are that ten sentences *be* revised.

CONDITIONS CONTRARY TO FACT

  If I *were* a little child, I would have no responsibilities.

  If I *were* you, I would not go.

  Had the weather *been* good, we would have gone to the top of the
  mountain.

In modern English the subjunctive survives mainly as a custom
in some expressions.

SUBJUNCTIVE

  The new manager requested that ten apartments *be* remodeled.

SUBJUNCTIVE NOT USED

  The new manager decided to have ten apartments remodeled.

■ Exercise 10

*Change the mood of the verbs to subjunctive when appropriate in
the following sentences. Put a C by those sentences that already
use the correct subjunctive.*

     were
1. If it was not for the ozone layer, ultraviolet rays could penetrate

    our atmosphere unhampered.

2. In the will he is drawing up, the billionaire requests that his
       be
    cats are well cared for after his death.

3. If you find that the vase is a copy rather than an original, you
                   be
    can demand that your money is refunded.

C   4. This house would be worth more on the current real estate

      market if it were not painted purple.

                                         were
5. People in many countries wish that the entire world was at

    peace.

7   Subject and Verb: Agreement   *agr*

Use singular verbs with singular subjects, plural verbs with plural subjects.

The *-s* or *-es* ending of the present tense of a verb in the third person *(she talks, he wishes)* indicates the singular. (The *-s* or *-es* ending for most *nouns* indicates the plural.)

SINGULAR	PLURAL
The *door* **opens**.	The *doors* **open**.
The *noise* **disturbs** the sleepers.	The *noises* **disturb** the sleepers.

7a   A compound subject (see p. 16) with *and* takes a plural verb.

> *Work and play* **are** not equally rewarding.
>
> *Golf and polo* **are** usually outdoor sports.

EXCEPTION: Compound subjects connected by *and* but expressing a singular idea take a singular verb.

> *The rise and fall* of waves **draws** a sailor back to the sea.
>
> When the children are in bed, *the tumult and shouting* **dies.**
>
> The *coach and history teacher* **is** Mr. Silvo.

7b   After a compound subject with *or, nor, either . . . or, neither . . . nor, not . . . but, not only . . . but also,* the verb agrees in number and person with the nearer part of the subject (see **8b**).

NUMBER

> Neither the *photographs* nor the *camera* **was** damaged by the fire.
>
> Either *fans* or an *air conditioner* **is** necessary.
>
> Either an *air conditioner* or *fans* **are** necessary.

PERSON

> Neither *you* nor your *successor* **is** affected by the new regulation.

7c   Intervening phrases or clauses do not affect the number of a verb.

Connectives like *along with* and *as well as* are not coordinating conjunctions but prepositions that take objects; they do not form compound subjects. Other such words and phrases include *in addition to, including, plus, together with, with.*

SINGULAR SUBJECT, INTERVENING PHRASE, SINGULAR VERB

The *pilot* as well as all his passengers *was* rescued.

Written with a coordinating conjunction, the sentence takes a plural verb.

The *pilot* **and** his *passengers were* rescued.

NOTE: Do not be confused by inversion.

From kind acts **grows** [not *grow*] *friendship.*

7d   A collective noun takes a singular verb when referring to a group as a unit, a plural verb when the members of a group are thought of individually.

A collective noun names a class or group: *congregation, family, flock, jury.* When the group is regarded as a unit, use the singular.

The *audience* at a concert sometimes **determines** the length of the performance.

When the group is regarded as separate individuals, use the plural.

The *audience* at a concert **vary** in their reactions to the music.

7e   Most nouns plural in form but singular in meaning take a singular verb.

Economics and *news* (and other words like *genetics, linguistics,* etc.) are considered singular.

Economics **is** often thought of as a science.

The *news* of a cure **is** encouraging.

Scissors and *trousers* are treated as plural except when used after *pair*.

> The *scissors* **are** dull.
>
> That *pair* of scissors **is** dull.
>
> The *trousers* **are** pressed and ready to wear.
>
> An old *pair* of trousers **is** sometimes stylish.

Other nouns that cause problems are *athletics*, *measles*, and *politics*. When in doubt, consult a dictionary.

7f   Indefinite pronouns *(anybody, anyone, each, either, everybody, everyone, neither, no one, nobody, one, some-body, someone)* usually take singular verbs.

> *Neither* of the explanations **was** satisfactory.
>
> *Everybody* **has** trouble choosing a subject for a paper.
>
> *Each* of the students **has** chosen a subject.

7g   Some words such as *all, some, none, part, half* (and other fractions) take a singular or a plural verb, depending on the noun or pronoun that follows.

> *singular*
>
> *Some* of the *sugar* **was** spilled on the floor.

> *plural*
>
> *Some* of the *apples* **were** spilled on the floor.

singular

Half of the *money* **is** yours.

plural

Half of the *students* **are** looking out the window.

When *none* can be regarded as either singular or plural, a singular or plural verb can be used.

plural   *singular*

None of the *roads* **are** closed. **OR** *None* of the *roads* **is** closed.

The number is usually singular.

singular

The number of *translations* **was** never determined.

A number when used to mean *some* is always plural.

plural

A number of translations of the phrase **were** suggested.

7h In sentences beginning with *there* or *here* followed by verb and subject, the verb is singular or plural, depending on the subject.

There and *here* (never subjects of a sentence) are sometimes **expletives** used when the subject follows the verb.

verb *subject*

There **was** a long *interval* between the two discoveries.

There **were** thirteen *blackbirds* perched on the fence.

Here **is** a *principle* to remember.

Here **are** two *principles* to remember.

The singular *is* may introduce a compound subject when the first noun is singular after an expletive.

There **is** *a swing and a footbridge* in the garden.

NOTE: In sentences beginning with *It,* the verb is always singular.

It **was** many years ago.

7i A verb agrees with its subject, not with a predicate nominative.

NO

His horse and *his dog* **are** his main source of pleasure.

NO

His main *source* of pleasure **is** his horse and his dog.

7j After a relative pronoun *(who, which, that),* the verb has the same person and number as the antecedent.

antecedent *relative pronoun* ⟶ *verb of relative pronoun*

Those who→**were** invited came.

We who→**are** about to die salute you.

The *costumes that*→**were** worn in the ballet were dazzling.

He was the *candidate who*→**was** able to carry out his pledges.

He was one of the *candidates who*→**were** able to carry out their pledges.

BUT

He was the only *one* of the candidates *who*→**was** able to carry out his pledges.

7k   A title or a word used as a word is singular and requires a singular verb even if it contains plural words and plural ideas.

Cats **is** an appealing musical.

"Hints from Heloise" **is** a syndicated newspaper column.

Oodles **is** a word that is still considered informal.

7L   Expressions of time, money, measurement, and so forth take a singular verb when the amount is considered a unit.

Two tons **is** a heavy load for a small truck.

Forty-eight hours **is** a long time to go without sleep.

■ Exercise 11

Correct any verb that does not agree with its subject.

1. The screech of seagulls mingle with the crash of waves on

   the beach.

2. In O'Neill's *Long Day's Journey into Night,* Mary's smiles and

    are
    laughter ~~is~~ increasingly forced, her resentment is more ob-

    vious, and her journey into night is more plainly marked.

3. *The Aspern Papers* deal**s** with the right of privacy.

4. A number of different careers are now open to those who

    major**s** in the humanities.

5. A book on statistical methods in the social sciences together

    is
    with several essays ~~are~~ required reading in the course.

6. Either *cacti* or *cactuses* **is** ~~are~~ acceptable for the plural form of

    cactus.

7. Molasses **was** ~~were~~ used in a great number of early New England

    recipes.

8. This tribal custom is enforced by strict taboos, the violation

    s
    of which bring ostracism.

9. Childish sentences or dull writing **is** ~~are~~ not improved by a

    sprinkling of dashes.

10. Neither the senator's son nor the senator's daughter wants a

    has
    career in politics, which ~~have~~ always fascinated their father.

8 Pronouns: Agreement, Reference, and Usage *agr/ref*

Use singular pronouns to refer to singular antecedents, plural pronouns to refer to plural antecedents. Make a pronoun refer to a definite antecedent.

The *debater* made **her** point eloquently.

The *debaters* made **their** points eloquently.

8a In general, use a plural pronoun to refer to a compound antecedent linked with *and*.

The *owner* and the *captain* refused to leave **their** distressed ship.

If two nouns designate the same person, the pronoun is singular.

The *owner and captain* refused to leave **his** distressed ship.

8b After a compound antecedent linked with *or, nor, either . . . or, neither . . . nor, not . . . but, not only . . . but also,* a pronoun agrees with the nearer part of the antecedent (see **7b**).

Neither the chess *pieces* nor the *board* had been placed in **its** proper position.

Neither the *board* nor the chess *pieces* had been placed in **their** proper positions.

A sentence like this written with *and* is less artificial and stilted.

The chess *pieces* **and** the *board* had not been placed in **their** proper positions.

8c A singular pronoun follows a collective noun antecedent when the members of the group are considered as a unit, but a plural pronoun follows when they are thought of individually (see **7d**).

A UNIT

The *committee* presented *its* report.

INDIVIDUALS

The *committee,* some of *them* smiling, filed into the room and took *their* seats.

8d Such singular antecedents as *each, either, neither, one, no one, everyone, someone, anyone, nobody, everybody, somebody, anybody* usually call for singular pronouns.

Not *one* of the linemen felt that **he** had played well.

Be consistent in number within the same sentence.

INCONSISTENT   *singular*   *plural*

Everyone takes **their** seats.

CONSISTENT

Everyone who is a mother sometimes wonders how **she** will survive the day.

Traditionally the pronouns *he* and *his* were used to refer to both men and women when the antecedent was unknown or representative of both sexes: "Each person has to face **his** own destiny." *He* was considered generic, that is, a common gender. Today this usage has changed because many feel that it ignores the presence and importance of women: there are as many "she's" as "he's" in the world, and one pronoun should not be selected to represent both.

Writers should be sensitive to this issue. Following are three alternatives to using the masculine pronoun:

1. Make the sentence plural.

   All *persons* have to face *their* own destinies.

2. Use *he or she* (or *his or her*).

   Each person has to face *his or her* own destiny.

3. Use *the* or avoid the singular pronoun altogether.

   Each person must face *the* future.
   Each person must face destiny.

8e   *Which* refers to animals and things. *Who* and *whom* refer to persons and sometimes to animals or things called by name. *That* refers to animals or things and sometimes to persons.

The *man* **who** was taking photographs is my uncle.

The *dog,* **which** sat beside him, looked happy.

Secretariat, **who** won the Kentucky Derby, will be remembered as one of the most beautiful horses of all time.

Sometimes *that* and *who* are interchangeable.

   A *mechanic that (who)* does good work stays busy.
   A *person that (who)* giggles is often revealing embarrassment.

NOTE: *Whose* (the possessive form of *who*) may be less awkward than *of which,* even in referring to animals and things.

The *car* **whose** right front tire blew out came to a stop.

8f Pronouns should not refer vaguely to an entire sentence or to a clause or to unidentified people.

VAGUE

> Some people worry about wakefulness but actually need little sleep. *This* is one reason they have so much trouble sleeping.

This could refer to the worry, to the need for little sleep, or to psychological problems or other traits that have not even been mentioned.

CLEAR

> Some people have trouble sleeping because they lie awake and worry about their inability to sleep.

It, them, they, and *you* are sometimes used as vague references to people and conditions that need more precise identification.

VAGUE

> *They* always get *you* in the end.

The problem here is that the pronouns *they* and *you* and the sentence are so vague that the writer could mean almost anything pessimistic. The sentence could refer to teachers, deans, government officials, or even all of life.

NOTE: At times writers let *this, which,* or *it* refer to the whole idea of an earlier clause or phrase when no misunderstanding is likely.

> The grumbler heard that his boss had called him incompetent. *This* made him resign.

8g Make a pronoun refer clearly to one antecedent, not uncertainly to two.

UNCERTAIN

    The agent visited her client before *she* went to the party.

CLEAR

    Before the client went to the party, the agent visited her.

8h Use pronouns ending in *-self* or *-selves* only in sentences that contain antecedents for the pronoun.

CORRECT INTENSIVE PRONOUN

    The cook *himself* washed the dishes.

FAULTY

    The antiques dealer sold the chair to my roommate and *myself*.

CORRECT

    The antiques dealer sold the chair to my roommate and *me*.

■ Exercise 12

Revise sentences that contain errors in usage of pronouns.

  1. The captain of the freighter let each member of the crew
             he or she
    decide whether ~~they~~ wished to remain with the ship.
                      me
  2. On behalf of my wife and ~~myself~~, I welcomed the visitors.
               its
  3. The green frog uses ~~his~~ tongue to catch insects.

4. The wire fence did not stop either the owner's dog or the
   its
   neighbor's cat from ~~their~~ efforts to get into the enclosure.

5. The drifter, along with his many irresponsible relatives, never
   he
   paid back a cent ~~they~~ borrowed.

6. In the early days in the West, almost every man could ride
   his
   ~~their~~ horse well.

   it
7. The League of Nations failed because ~~they~~ never received full

   support from member countries.

8. Was the first section of the essay written by your coauthor or
   you
   by ~~yourself~~?

   truth
9. The title of the new film is *Truth*, but ~~it~~ is not always easy to

   determine.

   she
10. None of the bridesmaids believed that ~~they~~ would have
    her
    occasion to wear ~~their~~ lavender dress again.

■ Exercise 13

*Revise sentences that contain vague or faulty references of pro-
nouns.*

   A
1. ~~In~~ ~~a~~n article in today's paper, ~~it~~ says that a cure has been

   found for laziness.

short-lived
2. There was some controversy about the duration of the meeting

involving heads of state, ~~but it did not last long.~~

Mints
3. ~~They~~ do not use as much silver in coins as they used to.

employers
4. The typical industrial worker is now well paid, but ~~they~~ have

not been able to do much about the boredom.

The government
         s
5. ~~They~~ tell you that you must pay taxes, but most of the time

        it   **s**
you do not know what ~~they~~ use your money for.

■ **Exercise 14**

Revise sentences that contain errors in reference of pronouns.
Write C by correct sentences.

who
1. That particular woman is the one ~~which~~ inherited millions of

dollars and gave it all away.

As usual in such a situation, the attorney advised the witness
2. ~~The advice of the attorney was that the witness refuse to~~
not to answer questions because of the possibility of
~~answer questions because of the possibility of self-~~
self-incrimination.
~~incrimination, which is usual in such a situation.~~

3. Cattle egrets are so named because they are frequently seen

       the cows
with cows, but ~~they~~ do not seem to mind them.

         she
4. Neither the mother nor her daughter knew that ~~they~~ had a

wealthy relative.

5. On the night of July 14, the patriots stormed the doors of the
   them
   jail, and ~~they were~~ immediately smashed ∧ open.
                                        who
6. The lecturer ignored the heckler, ~~because he~~ was obviously

   unwell.

7. The poet is widely read, but it is very difficult indeed to make
   by writing poetry.
   a living ∧ ~~at it.~~
                                              who are not unusually poor
8. Lawyers generally charge their clients ∧ a standard fee. ~~unless~~

   ~~they are unusually poor.~~

C 9. The osprey feeds on fish, which it captures by diving into the

   water.
                        his or her
10. Everyone should do ~~their~~ part to make the world a little

   better.

9 Case   *c*

Use correct case forms.

Case expresses the relationship of pronouns *(me, I)* and nouns to
other words in the sentence by the use of different forms. Nouns
are changed in form only for the possessive case *(child, child's;*
see p. 473).

Following is a chart of the cases of pronouns:

PERSONAL PRONOUNS

Singular	*Subjective*	*Possessive*	*Objective*
First person	I	my, mine	me
Second person	you	your, yours	you
Third person	he, she, it	his, her, hers, its	him, her, it

Plural	*Subjective*	*Possessive*	*Objective*
First person	we	our, ours	us
Second person	you	your, yours	you
Third person	they	their, theirs	them

RELATIVE OR INTERROGATIVE PRONOUNS

Number	*Subjective*	*Possessive*	*Objective*
Singular	who	whose	whom
Plural	who	whose	whom

To determine case, decide how a word is used in its own clause—for example, whether it is a subject or a subjective complement or an object.

9a   Use the subjective case for subjects and for subjective complements that follow linking verbs.

SUBJECT

After seven years, my former *roommate* and *I* [not *me*] had much to talk about.

It looked as if my *friend* and *I* [not *me*] were going to be roommates again.

SUBJECTIVE COMPLEMENT

The fortunate ones were *you* and *I.*

In speech, *it's me, it's us, it's him,* and *it's her* are sometimes used. These forms are not appropriate for formal writing.

9b Use the objective case for a direct object, an indirect object, or the object of a preposition.

DIRECT OBJECT
The magician *amazed* Yolanda and **me** [not I].

INDIRECT OBJECT
Mozart *gave* **us** great music.

OBJECT OF PREPOSITION
The director had to choose *between* **her** and **me** [not she and I].

NOTE: Be careful about the case of pronouns in constructions like the following:

FAULTY
A few *of* **we campers** learned to cook.

CORRECT
A few *of* **us campers** learned to cook.

When in doubt, test by dropping the noun:

A few *of* **us** learned to cook.

CORRECT (when pronoun is subject)
We campers learned to cook.

9c Use the objective case for subjects and objects of infinitives.

subject of infinitive
The editor considered **her** *to be* the best reporter on the staff.

9d Use the same case for an appositive and the word to which it refers (see 23b).

The case of a pronoun appositive depends on the case of the word it refers to.

SUBJECTIVE
> *Two delegates*—Esther Giner and **I**—were appointed by the president.

OBJECTIVE
> The president appointed two *delegates*—Esther Giner and **me**.

9e The case of a pronoun after *than* or *as* in an elliptical (incomplete) clause should be the same as if the clause were completely expressed.

subject of understood verb

No one else in the play was as versatile as **she** *(was)*.

object of understood subject and verb

Her fellow actors respected no one more than *(they respected)* **her**.

9f Use the apostrophe or an *of* phrase to indicate the possessive case (see 33).

> The club*'s* motto was written in Latin.
> The motto *of the club* was written in Latin.

9g Use the possessive case for pronouns and nouns preceding a gerund.

gerund
↓
None of my friends had heard about **my** [not **me**] *leaving*.

gerund
↓
The **lumberman's** *chopping* could be heard for a mile.

BUT
A noun before a gerund (see p. 22) may be objective when a **phrase** intervenes:

The *baby* **in the next apartment** *crying* kept us awake.

when the noun is **plural**:

There is no rule against **people** *working* overtime.

when the noun is **abstract**:

I object to **emotion** *overruling* judgment.

when the noun denotes an **inanimate object**:

The police officer did not like the **car** *being* parked in the street.

When a verbal is a participle and not a gerund, a noun or pronoun preceding it is in the objective case. The verbal functions as an adjective.

No one saw **anyone** *running* from the scene.

9h The possessive forms of personal pronouns have **no** apostrophe; the possessive forms of indefinite pronouns **do have** an apostrophe.

PERSONAL PRONOUNS
yours   its   hers   his   ours   theirs

INDEFINITE PRONOUNS
everyone's   other's   one's   anybody else's

9i The case of an interrogative or a relative pronoun is determined by its use in its own clause.

Interrogative pronouns (used in questions) and relative pronouns are *who, whose, whom, what, which. Who* is used for the subjective case; *whom,* for the objective case.

In formal writing always use *whom* for objects.

The hostess did not tell her family **whom** she *had invited* to dinner.

In speech *who* is usually the form used at the beginning of a sentence, especially an interrogative sentence.

Who was your mother talking to?

The case of pronouns is clear in brief sentences.

Who *conceived* the idea of a committee?

But when words intervene between the pronoun and the main verb, determining the case can often be difficult.

Who do they say *conceived* the idea of a committee?

Mentally cancel the intervening words.

> **Who** ~~do they say~~ *conceived* the idea of a committee?

Do not confuse the function of the relative pronoun in its clause with the function of the clause as a whole. Pick out the relative clause and draw a box around it. Then the use of the pronoun in the dependent clause is more easily determined.

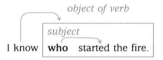

object of verb

subject

I know **who** started the fire.

Try to avoid writing sentences with elaborate clauses that make the choice between *who* and *whom* difficult.

■ Exercise 15

Underline the correct word in each of the following sentences.

1. (<u>Who</u>, Whom) do historians believe built the first steam engine?

2. In Greek mythology the Muses are nine daughters of Zeus, each of (who, <u>whom</u>) is supposed to reign over a different art or science.

3. An education is there for (<u>whoever</u>, whomever) will take advantage of the opportunity.

4. Will the delegate from the Virgin Islands please indicate (<u>who</u>, whom) is to receive her delegation's votes?

5. The mission gives (<u>whoever</u>, whomever) comes a hot meal and a secure place to spend the night.

6. (<u>Whoever</u>, Whomever) is elected will have to deal with the problem.

7. There was in those days in Paris a singer (<u>who</u>, whom) the secret police knew was a double agent.

8. On the platform stood the man (<u>who</u>, whom) they all believed had practiced witchcraft.

9. On the platform stood the man (who, <u>whom</u>) they all accused of practicing witchcraft.

10. The speaker defended his right to talk critically of (whoever, <u>whomever</u>) he pleased.

■ Exercise 16

Cross out the incorrect forms of pronouns and nouns, and write in the correct forms.

                                          she   **I**
1. No one enjoyed milkshakes more than ~~her~~ unless it was ~~me.~~

2. It was plain to ~~we~~ **us** students that the professor was delivering the same lecture that we had heard the week before.

3. If the ordinance were passed granting a right of way to you and ~~I~~ **me**, whose responsibility would it be to make sure that we are allowed access?

4. Between ~~she~~ **her** and ~~I~~ **me** there never were any secrets; at least, that is what she told me.

5. I decided to allow nothing to interfere with ~~me~~ **my** studying.

6. No one was able to make more intricate designs than ~~him~~ **he**.

7. Just between you and ~~I~~ **me**, ~~whom~~ **who** is it that the judges are favoring?

8. Evan and ~~him~~ **he** appeared just before sunset in that old car, waving and shouting at Eileen and me.

9. The physician said that he had not objected to the employee**'s** returning to work.

10. After much discussion between the Navajo and ~~she~~ **her**, they agreed that the first chance to buy the turquoise bracelet was ~~her's~~ **hers** rather than the man's.

10   Adjectives and Adverbs   *adj/adv*

Use adjectives to modify nouns and pronouns, adverbs to modify verbs, adjectives, and other adverbs.

Most adverbs end in *-ly*. Only a few adjectives (such as *lovely, holy, manly, friendly*) have this ending. Some adverbs have two forms, one with *-ly* and one without: *slow* and *slowly, loud* and *loudly*. Most adverbs are formed by adding *-ly* to adjectives: *warm, warmly; pretty, prettily*.

Choosing correct adjectives and adverbs in some sentences is simple.

> They stood *close*. [adverb]
> She had a *close* relative in St. Louis. [adjective]

Adjectives do not modify verbs, adverbs, or other adjectives. Distinguish between *sure* (adjective) and *surely* (adverb); *easy* and *easily; good* and *well; real* and *really*.

NOT

> Balloonists *soar* over long distances **easy**. [adjective]

BUT

> Balloonists *soar* over long distances **easily**. [adverb]

10a   Use the comparative to refer to two things, the superlative to more than two.

> *Both* cars are fast, but the small car is (the) **faster**.
> All *three* cars are fast, but the small car is (the) **fastest**.

10b Add *-er* and *-est* to form the comparative and superlative degrees of most short modifiers. Use *more* and *most* (or *less* and *least*) before long modifiers.

ADJECTIVES	COMPARATIVE	SUPERLATIVE
	-er/-est	
dear	dearer	dearest
pretty	prettier	prettiest
	more/most	
pitiful	more pitiful	most pitiful
grasping	more grasping	most grasping
	less/least	
expensive	less expensive	least expensive
ADVERBS		
	-er/-est	
slow	slower	slowest
	more/most	
rapidly	more rapidly	most rapidly
	less/least	
sensitively	less sensitively	least sensitively

Some adjectives and adverbs have irregular forms: *good, better, best; well, better, best; little, less, least; bad, worse, worst.* Consult a dictionary.

Some adjectives and adverbs are absolute; that is, they cannot be compared *(dead, perfect, unique).* A thing cannot be more or less dead or perfect or unique (one of a kind). Acceptable forms are *more nearly perfect* or *almost dead.*

10c Use a predicate adjective, not an adverb, after a linking verb (see p. 7) such as *appear, be, become, feel, look, seem, smell, sound, taste.*

ADJECTIVE

He feels **bad.** [He is ill or depressed. An adjective modifies a pronoun.]

ADVERB

He *reads* **badly.** [*Reads* is not a linking verb. An adverb modifies a verb.]

ADJECTIVE

The *tea* tasted **sweet.** [*Sweet* describes the tea.]

ADVERB

She *tasted* the tea **daintily.** [*Daintily* tells how she tasted the tea.]

10d Use an adjective, not an adverb, to follow a verb and its object when the modifier refers to the object, not to the verb.

Verbs like *keep, build, hold, dig, make, think* are followed by a direct object and a modifier. After verbs of this kind, choose the adjective or the adverb form carefully.

ADJECTIVES—MODIFY OBJECTS

Keep your *clothes* **neat.**

Make the *line* **straight.**

ADVERBS—MODIFY VERBS

> *Arrange* your clothes **neatly** in the closet.

> *Draw* the line **carefully**.

■ Exercise 17

Underline unacceptable forms of adjectives and adverbs, and write the correct forms. If a sentence is correct, write C.

1. Some modern styles of clothing look <u>similarly</u> **[similar]** to those of

   many years ago.

2. Workers outside the building made so much noise that we
   could <u>scarce</u> **[scarcely]** hear the speaker, though his shirt <u>sure</u> **[surely]** was loud.

3. Nevertheless, we felt <u>badly</u> **[bad]** because we had to leave the

   lecture before it was over.

4. Adelaide Crapsey was not one of the most <u>popularest</u> **[popular]** poets

   of her time, but some people believe that she was one of the

   most talented.

5. Hungry birds strip a holly bush of its berries <u>rapid</u> **[rapidly]** and swoop
   away seeking <u>frantic</u> **[frantically]** for other food.

6. The computer, a <u>real</u> **[really]** complicated mechanical mind, is <u>the</u> **[a]**
   most <u>unique</u> **[unique]** instrument of modern civilization.

C   7. The athlete played awkwardly and badly but was the better

of the two choices.

            extremely   **really**

8. The stockholders were <u>mighty</u> unhappy but not <u>real</u> surprised

when the corporation lost money for another year.

           logically

9. In times of tribulation, a leader must think <u>logical</u>.

10. The manager never knew which of the two sales programs

          more

was likely to be the <u>most</u> successful.

Sentence
Structure

81

11 Choppy Sentences and Excessive Coordination *chop/coord*

Do not string together brief independent clauses or short sentences (see 19).

Wordiness and monotony result from choppy sentences or from strings of brief independent clauses connected by coordinating conjunctions *(and, but, or, nor, for, yet, so)*. Furthermore, excessive coordination does not show precise relationships between thoughts. Skillful writers vary their sentences by using phrases and dependent clauses that demonstrate the logical connections between ideas.

STRINGY

> The Valkyries appear in Norse poetry, and they are large women, and they have beautiful skin, and they are golden haired, and they wear helmets and breastplates, and they carry spears and shields, and they ride their horses through the sky.

CHOPPY

> The Valkyries appear in Norse poetry. They are large women. They have beautiful skin. They have golden hair. They wear helmets and breastplates. They carry spears and shields. They ride their horses through the sky.

IMPROVED

> The Valkyries of Norse poetry are large women with beautiful skin and golden hair who wear helmets and breastplates and who carry spears and shields as they ride their horses through the sky.

■ Exercise 1

Improve the following sentences by subordinating some of the ideas. Combine choppy sentences into longer ones.

1. ~~The automobile was not the invention of any one person. No~~ **Evolving from two hundred years of experimentation in which French, Americans, British, Germans, and others were involved, the automobile was not the invention of any one person or nation.** ~~single nation can be given credit for its development. It evolved over time. It resulted from two hundred years of experimentation. French, Americans, British, Germans, and others were involved.~~

2. Samuel Johnson**,** ~~was~~ a famous British writer of the eighteenth **who** century**,** ~~and he~~ was raised in poverty, ~~and he~~ once said,

   "Slow rises worth by poverty oppressed."

3. ~~In early times, coins had intrinsic value. They were worth~~ **In early times when coins had intrinsic value and were worth whatever the metal in them was worth at that moment,** ~~whatever the metal in them was worth at that moment. They~~ **merchants and bankers were kept busy constantly** ~~had to be weighed. Merchants and bankers were kept busy~~ **determining the value of coins by weighing them.** ~~weighing coins. They had to be constantly determining the value of coins.~~

4. ~~The point-contact transistor was invented in 1948. It was~~ **Two scientists who worked at the Bell Telephone Laboratories, John Bardeen and Walter H. Brattain, invented the point-contact transistor in 1948.** ~~invented by two scientists. Their names are John Bardeen and Walter H. Brattain. They worked at the Bell Telephone Laboratories.~~

5. The manta ray has a wide, flat body, ~~and it is also called a~~ *graceful* , *or devilfish,* .

   ~~devilfish, and it is graceful.~~

6. ~~Pagodas are~~ temples or ~~sacred buildings, and they~~ are found *Sacred* , *pagodas, which*

   in several Eastern countries, ~~and they~~ often have many stories

   and upward-curving roofs.

7. ~~Sharks are~~ ferocious, ~~And they~~ attack many bathers each *Although* *sharks*

   year, ~~but~~ they seldom kill, so their reputation as killers is in

   part undeserved.

8. Some vacationers leave home in search of quiet, ~~so they~~ find *and*

   a place without a telephone or television, but other people *who*

   want complete isolation, ~~but they discover that it is difficult~~ *often cannot*

   ~~to~~ find a park that is not crowded with trailers and tents.

9. The instructor gave the new student an assignment, ~~and he~~ —

   ~~had~~ to write just one sentence, but he could not think of an *;* *because*

   interesting subject, ~~and so~~ he did not do the required work.

10. Benjamin Franklin ~~was~~ an American, ~~He~~ was at home wherever *,* *who*

    he went, ~~He~~ gained wide popularity in France, ~~He was~~ also *and*

    well known in England. *became*

12 Subordination   *sub*

Use subordination to achieve proper emphasis and effective, varied construction.

Putting the less important idea of a sentence in a dependent clause emphasizes the more important thought in the independent clause. However, piling one dependent clause on top of another awkwardly stretches out sentences and obscures meaning.

12a Express main ideas in independent clauses, less important ideas in dependent clauses.

An optimistic sociologist wishing to stress progress despite crime would write:

> Although the crime rate is high, society has progressed in some ways.

A pessimistic sociologist might wish the opposite emphasis:

> Although society has progressed in some ways, the crime rate is high.

Upside-down subordination results from placing the main idea of a sentence in a dependent clause. The writer of the following sentence intended to stress the insight and wisdom possessed by a young woman despite her youth.

> ┌──────────── *dependent clause* ────────────┐ ┌*independent*
> Although she possessed unusual insight and wisdom, she was still
>
> *clause*┐
> young.

Unintentionally the writer of the above sentence emphasized the person's youth. For the proper stress, the sentence should read:

┌─────*dependent clause*─────┐ ┌─────── *independent clause* ───────
Although she was still young, she possessed unusual insight and

┌──────────┐
wisdom.

12b Avoid excessive overlapping of dependent clauses.

Monotony and even confusion can result from a series of clauses in which each depends on the previous one.

OVERLAPPING

> Pianos are instruments
>
>   that contain metal strings
>
>   that make sounds when struck by felt-covered hammers
>
>   that are operated by keys.

IMPROVED

> Pianos are instruments containing metal strings that make sounds when struck by felt-covered hammers operated by keys.

■ Exercise 2

The following is an exercise in thinking about relationships. The sentences used are designed to point out differences in meaning that result from subordination. Read the pairs of sentences carefully, and answer the questions.

1. A. When no one complains, the status quo is maintained.
   B. When the status quo is maintained, no one complains.

   Which sentence is likely to be written by a person who resists change? **B**
   Which seems to suggest the need for change? **A**

2. A. Even though the article was long and dull, the author made some interesting comments.
   B. Even though the author made some interesting comments, the article was long and dull.

   Which sentence reflects a more positive attitude toward the article? **A**

3. A. Although a lifetime is short, much can be accomplished.
   B. Although much can be accomplished, a lifetime is short.

   Which of these sentences expresses more determination? **A**

4. A. When in doubt, some drivers apply the brakes.
   B. When some drivers apply the brakes, they are in doubt.

   With which drivers would you prefer to ride? **A**

5. A. While taking a bath, Archimedes formulated an important principle in physics.
   B. While formulating an important principle in physics, Archimedes took a bath.

   Which sentence indicates accidental discovery? **A**
   In which sentence does Archimedes take a bath for relaxation? **B**

■ Exercise 3

Rewrite the following sentences to avoid excessive subordination.

1. ~~In a discovery that took place~~ I̲n 1939, construction workers
   discovered
   ~~found~~ a cave ~~that is~~ near Monte Circeo, Italy, that had been

   sealed for fifty thousand years.

2. Inside the cave ~~was~~ a circle ~~that was made~~ of rocks ~~that~~
   the broken
   surrounded ~~a~~ skull/ ~~which had been broken, that was that~~ of

   an ancient man.

3. Lobster Newburg is ~~a dish that consists of~~ cooked lobster
   with
   meat/ ~~which is~~ heated in a chafing dish ~~that contains~~ a special

   cream sauce.

   Jethro Tull
4. ~~Jethro Tull was an~~ English agriculturist, ~~who~~ is famous for
   wasteful
   departing from the old method of cultivation that involved

   planting seeds broadcast/ ~~which was wasteful.~~

   an economic and efficient
5. Tull developed a system/ ~~which was economic and efficient,~~
   his invention,
   of planting in rows ~~that were~~ created by a ~~machine that was~~

   ~~called~~ a seed drill that ~~he invented that~~ could dig a hole and

   drop in a seed in one operation.

13 Completeness *compl*

Make your sentences complete in structure and thought.

Every element of a sentence should be expressed or implied clearly to prevent inconsistency and misunderstanding.

13a Do not omit a verb or a preposition that is necessary to the structure of the sentence.

OMITTED PREPOSITION

> The apes were both attracted and suspicious **of** the stuffed animal placed in their midst.

IMPROVED

> The apes were both attracted **to** and suspicious **of** the stuffed animal placed in their midst.

BETTER

> The apes were both attracted **to** the stuffed animal placed in their midst and suspicious **of** it.

OMITTED VERB

> The cabinet *drawers* **were** open and the glass *door* shattered. [The door **were** shattered?]

VERB STATED

> The cabinet *drawers* **were** open, and the glass *door* **was** shattered.

When the same verb form is called for in both elements, it need not be repeated:

> To err is human; to forgive, divine.

13b Omission of *that* sometimes obscures meaning.

INCOMPLETE

The labor leader reported a strike was not likely.

COMPLETE

The labor leader reported **that** a strike was not likely.

14 Comparisons   *comp*

Make comparisons logical and clear.

14a Compare only similar terms.

The *laughter* of a loon is more frightening than an **owl.**

This sentence compares a sound and a bird. A consistent sentence would compare sound and sound or bird and bird.

The *laughter* of a loon is more frightening than the **hoot** of an owl.

A *loon* is more frightening than an **owl.**

14b The word *other* is often needed in a comparison.

INCOMPLETE

The Ganges is longer than any river in India.

COMPLETE (when *other* added)

The Ganges is longer than any *other* river in India.

14c   Avoid awkward and incomplete comparisons.

AWKWARD AND INCOMPLETE

The new cars appear to be as small if not smaller than last year's models. [*As small* requires *as,* not *than.*]

BETTER

The new cars appear to be as small as last year's models if not smaller.

AWKWARD AND INCOMPLETE

During much of the nineteenth century, Queen Victoria of Great Britain was one of the best known if not the best known woman in the world. [After *one of the best known,* the plural *women* is required.]

BETTER

During much of the nineteenth century, Queen Victoria of Great Britain was one of the best-known women in the world if not the best known.

AMBIGUOUS

After many years my teacher remembered me better than my roommate. [Better than he remembered my roommate, or better than my roommate remembered me?]

CLEAR

After many years my teacher remembered me better than my roommate did.

OR

After many years my teacher remembered me better than he did my roommate.

■ Exercise 4

Correct any errors in completeness and comparisons. Write C by correct sentences.

                    other
1. The aorta is larger than any⌄artery of the body.

                                     as
2. Common sense is valued among many people as highly ~~if not~~
                    , if not higher.
   ~~higher than~~ intellectualism/⌄

                      does *or* politics does.
3. Diplomacy calls for more tact than⌄politics.

                      in
4. Taxes are usually higher in urban areas than⌄the country.

                       as
5. For good health, plain water is as good ~~if not better than~~ most
             , if not better.
   other liquids/⌄

                      of
6. The lighthouse stood as a symbol⌄and guide to safety.

C 7. Some laws are so broad that they allow almost unlimited

     interpretations.

                            did.
8. The baboons ate more of the bananas than the ants/⌄

9. Statistics reveal that the average age of our nation's populace
      increasing
   is⌄and will continue to increase.

10. The veterinarian read an article contending that horses like
            they do *or* cats do.
   dogs better than⌄cats.

■ Exercise 5

Follow the instructions for Exercise 4.

1. Auckland is larger than any **other** city in New Zealand.

2. Copernicus discovered **that** the sun is at the center of our solar

   system.

3. Birds seem to sing more beautifully in the early morning than

   in the afternoon.

4. The editors say that the headlines have been written and the

   type **has been** set.

5. Shoppers standing in the checkout line are sometimes both

   curious **about *or* about the tabloids on the display racks and** and disapproving of the tabloids on the display racks.
   disapproving of them.

15 Consistency   *cons*

Write sentences that maintain consistency in form and meaning.

15a Avoid confusing or awkward shifts in grammatical forms.

Tenses

PRESENT AND PAST
SHIFT

> The actors *rehearsed* and rehearsed, and then they finally *perform*.
> [Use *rehearsed . . . performed* or *rehearse . . . perform*.]

CONDITIONAL FORMS *(should, would, could)*
SHIFT

> Exhaustion after a vacation *could* be avoided if a family *can* plan
> better. [Use *could . . . would* or *can . . . can*.]

Person

> In department stores, *good salespersons* [**3rd person**] first ask whether
> a customer needs help or has questions. Then *they* [**3rd person**] do
> not hover around. If *you* [**2nd person**] do, *you* [**2nd person**] run the
> risk of making the customer uncomfortable or even angry. [The third
> sentence should read, "If they do, they run . . ."]

Number

> A *witness* may see an accident one way when it happens, and then
> *they* remember it an entirely different way when *they* testify. [Use
> *witness* and a singular pronoun or *witnesses* and *they . . . they*.]

Mood

SHIFT

>   *indicative*   *imperative*
>   ↓   ↓
> First the job-seeker *mails* an application; then *go* for an interview.

CONSISTENT

> First the job-seeker *mails* an application; then he or she *goes* for an
> interview.
>
> First *mail* an application; then *go* for an interview.

Voice

SHIFT

> The chef *cooks* [**active**] the shrimp casserole for thirty minutes, and then it *is allowed* [**passive**] to cool. [Use *cooks* and *allows*. See **5**.]

Connectors

RELATIVE PRONOUN

> She went to the chest of drawers *that* leaned perilously forward and *which* always resisted every attempt to open it. [Use *that . . . that*.]

CONJUNCTIONS

> The guests came *since* the food was good and *because* the music was soothing. [Use *since . . . since* or *because . . . because*.]

Direct and indirect discourse

MIXED

> The swimmer says that the sea is calm and why would anyone fear it?

CONSISTENT

> The swimmer says, "The sea is calm. Why would anyone fear it?"
>
> The swimmer says that the sea is calm and asks why anyone would fear it.

■ Exercise 6

Correct the shifts in grammar in the following sentences.

1. Halley's Comet ~~is~~ **was** first observed in ancient times and then

   was seen at regular intervals afterwards.

2. Success in the experiments will be achieved this time if the

   assistants ~~would~~ **will** follow the instructions precisely.

                             that

3. Contact lenses that are well fitted and ~~which~~ are well cared

    for usually cause few problems.

4. Ulysses S. Grant was looking forward to the day of Lee's

                           came

    surrender, but when that day ~~comes~~, he experienced a

                      suffered

    migraine headache and ~~suffers~~ through the ceremony.

5. The smiling actor backed carefully down the steps, jumped

                rode

    on his horse, and ~~rides~~ off, leaving a cloud of dust behind.

6. Enthusiastic runners like to get out every day, but occasionally

                        is

    the weather in this region ~~would be~~ too bad even for the most

    dedicated.

7. The losing candidate demanded a recount, and the judge

    ordered

    ~~orders~~ the ballots to be secured.

                        one *or* **he or she**

8. Before one purchases an electrical appliance, ~~you~~ should find

    out how expensive it will be to operate.

    Start   **use**

9. ~~It is wise to start~~ a fire with a wood like pine, and then ⋀ a

    heavier wood like oak.~~is used~~.

                    because

10. Use pine first to start a fire ~~since~~ it ignites easily and because

    it will then make the oak burn.

15b Make subjects and predicates fit logically together.
Avoid faulty predication.

Test a sentence by mentally placing the subject and predicate
side by side. If they do not fit together (**faulty predication**), rewrite
for clarity and logic.

FAULTY PREDICATION

Tragedy, according to Aristotle, *is when* a person of high estate falls.
[*Tragedy . . . is when* makes no sense.]

LOGICAL

Tragedy, according to Aristotle, *occurs when* a person of high estate
falls.

FAULTY

After *eating* a large meal is a bad *time* to go swimming.

LOGICAL

The *period* immediately after eating a large meal is a bad *time* to go
swimming.

FAULTY

The *use* of a mediator *was hired* so that a compromise could be
achieved.

LOGICAL

A *mediator was hired* so that a compromise could be achieved.

■ Exercise 7

Correct faulty predication in the following sentences.

              occurs
1. A free election ~~is~~ when voters are allowed to select the

candidate of their choice.

              the reason
2. Volcanic action is ‸ why many islands in the South Seas have

beaches with dark sand.

            C   **regarded as**
3. ~~The question of~~ ¢ensorship is ~~the answer~~ often ~~given to the~~
the solution to the problem of objectionable language in
~~objectionable language of~~ many modern novels.

         A
4. ~~By way of~~ ⱥ tunnel under the English Channel ~~it~~ could greatly

shorten the time it takes to travel from England to France.

         P   **only when**
5. ~~The only time~~ ₱eople should be tardy ~~is if~~ they cannot possibly

avoid being late.

16 Position of Modifiers   *dg/mod*

Attach modifiers clearly to the right word or element in the
sentence.

A misplaced modifier can cause confusion or misunderstanding.
Usually a modifying adjective precedes its noun whereas an adverb
may precede or follow the word it modifies. Prepositional phrases

usually follow closely but may precede. Adjective clauses follow closely. Adverbial phrases and clauses may be placed in many positions. (See pp. 21–25.)

16a Avoid dangling modifiers.

A verbal phrase at the beginning of a sentence should modify the subject.

DANGLING PARTICIPLE

Speeding down the hill, several slalom *poles* were knocked down by the skier.

CLEAR

Speeding down the hill, the *skier* knocked down several slalom poles.

DANGLING GERUND

After **searching** around the attic, a *Halloween mask* was discovered. [The passive voice in the main clause causes the modifier to attach wrongly to the subject.]

CLEAR

After **searching** around the attic, *I* discovered a Halloween mask.

DANGLING INFINITIVE

To **enter** the house, the *lock* on the back door was picked. [*To enter the house* refers to no word in this sentence.]

CLEAR

To **enter** the house, *he* picked the lock on the back door.

DANGLING PREPOSITIONAL PHRASE

After completing my household chores, the *dog* was fed.

CLEAR

After completing my household chores, *I* fed the dog.

DANGLING ELLIPTICAL CLAUSE

While still sleepy and tired, the *counselor* lectured me on breaking rules.

CLEAR

While I was still sleepy and tired, the counselor lectured me on breaking rules.

NOTE: Loosely attaching a verbal phrase to the end of a sentence is unemphatic.

UNEMPHATIC

Cultivated tomatoes are a relatively recent addition to our tables, having been widely used only in the last hundred years or so.

BETTER

Cultivated tomatoes are a relatively recent addition to our tables; they have been widely used only in the last hundred years or so.

Some verbal phrases that are **sentence modifiers** do not need to refer to a single word:

Strictly speaking, does this sentence contain a dangling construction?
To tell the truth, it does not.

16b Avoid misplaced modifiers.

Placement of a modifier in a sentence affects meaning.

He enlisted after he married *again*.
He enlisted *again* after he married.

Almost anything that comes between an adjective clause and the word it modifies can cause confusion.

MISLEADING

Even on a *meager* sheriff's salary, a family may have some comforts.

CLEAR

Even on a sheriff's *meager* salary, a family may have some comforts.

PUZZLING

A hungry tarantula seizes any insect that touches its hair *swiftly*.

CLEAR

A hungry tarantula *swiftly* seizes any insect that touches its hair.

16c A modifier placed between two words so that it can modify either word is said to squint.

SQUINTING   *— or —*

The secretary who was typing *slowly* rose from her chair.

CLEAR

The secretary who was *slowly typing* rose from her chair.

OR

The secretary who was typing *rose slowly* from her chair.

■ Exercise 8

Correct the faulty modifiers in the following sentences.

, while still in his pajamas,

1. Early one morning in Kenya, the anthropologist saw near the

   camp a white elephant. ~~while still in his pajamas~~.

      T

2. ~~Looking~~ /hrough a magnifying glass, the flaw in the diamond

   appeared as a dark spot.

      who have strange beliefs

3. Listeners ⋀ often call in to radio talk shows. ~~that have strange~~

   ~~beliefs.~~

      with sophisticated programs   **problems**

4. Computers ⋀ can solve ~~problems~~ in minutes ⋀ ~~with sophisticated~~

   ~~programs~~ that used to take months.

   Check the answer t **. *or* You must check the answer . . .**

5. ⋀ ⱦo be absolutely certain ⋀ ~~the answer must be checked.~~

      inexpensive

6. The restaurant offers ⋀ meals for children. ~~that are inexpensive~~.

7. Serve one of the melons for dessert at lunch; keep one of

                 for the picnic.

   them ~~for the picnic~~ in the refrigerator, / ⋀

   Before sawing, t

8. ⋀ Ⱦhe carpenter inspected the board ~~before sawing~~ for nails.

           T   **slowly and relentlessly**

9. ~~Slowly and relentlessly~~ Ⱦhe lecturer said that witches ⋀ attempt

   to gain control over the minds of others.

   Frequently t *or* frequently

10. ⋀ Ⱦhose who ⋀ lose sleep ~~frequently~~ cannot function properly.

■ Exercise 9

Follow the instructions for Exercise 8.

                  only

1. The least expensive electric clocks now ~~only~~ cost ⋀ about five

   or six dollars.

the passengers signed

2. Before landing at Plymouth, ∧the Mayflower Compact. ~~was~~

   ~~signed by the passengers~~.

the hikers found

3. Without shoes∧the rocky terrain ~~was~~ hard on the feet. ~~of the~~

   ~~hikers~~.

the divers entered

4. Wishing to prevent the bends, ∧the decompression chambers.

   ~~were entered by the divers~~.

                              T
5. ~~To taste delicious,~~ ∧the chef should prepare a dressing suitable

   make   **taste delicious.**
   to∧the raw spinach salad, ∧

17 Separation of Elements   *sep*

Do not needlessly separate closely related elements.

Separation of subject and verb, parts of a verb phrase, or verb
and object can result in awkwardness or confusion.

AWKWARD

   The Alamo is known because it symbolizes the struggle to break
   Mexican rule as "the cradle of Texas liberty."

IMPROVED

   The Alamo is known as "the cradle of Texas liberty" because it
   symbolizes the struggle to break Mexican rule.

PUZZLING

   She is the man who owns the service station's wife.

CLEAR

> She is the wife of the man who owns the service station.

Do not divide a sentence with a quotation long enough to cause excessive separation.

AVOID

> Stephen Clarkson's opinion that "politicians always create their own reality; during campaigns they create their own unreality" is pessimistic.

BETTER

> Stephen Clarkson's opinion is pessimistic: "Politicians always create their own reality; during campaigns they create their own unreality."

Split infinitives occur when a modifier comes between *to* and the verb form, as in *to loudly cheer*. Some writers avoid them without exception; others accept them occasionally. To avoid objections, do not use this kind of split construction.

18 Parallelism *paral*

Use parallel grammatical forms to express parallel thoughts.

Elements in a sentence are *parallel* when one construction matches another construction: a phrase and a phrase, a clause and a clause, a verb and a verb, a noun and a noun, a verbal and a verbal, and so forth.

18a Use parallel constructions with coordinating conjunctions (*and, but, for,* etc.).

NOT PARALLEL

          adjective   *verb*
          ↓   ↓

Sailing ships were *stately* and *made* little noise.

PARALLEL

           adjectives
        ↙   ↘

Sailing ships were *stately* and *quiet*.

NOT PARALLEL

          nouns   *pronoun*
        ↙   ↘   ↓

Young Lincoln read widely for *understanding, knowledge,* and *he* just
liked books.

PARALLEL

             nouns
             ↓

Young Lincoln read widely for *understanding, knowledge,* and *pleasure*.

NOTE: Repeat an article *(the, a, an)*, a preposition *(by, in, on,* etc.), the sign of the infinitive *(to)*, or other key words to clarify parallelism.

UNCLEAR

The artist was *a* painter and sculptor of marble.

CLEAR

The artist was *a* painter and *a* sculptor of marble.

UNCLEAR

They passed the evening *by* eating and observing the crowds.

CLEAR

They passed the evening *by* eating and *by* observing the crowds.

18b Use parallel constructions with correlatives (*not only . . . but also, either . . . or,* etc.).

NOT PARALLEL

infinitive   *preposition*

Petroleum is used **not only** *to make* fuels **but also** *in* plastics.

NOT PARALLEL

verb   *preposition*

Not only *is* petroleum used in fuels **but also** *in* plastics.

PARALLEL

prepositions

Petroleum is used **not only** *in* fuels **but also** *in* plastics.

NOT PARALLEL

adverb   *pronoun*

The speeches were **either** *too* long, **or** *they* were not long enough.

NOT PARALLEL

article   *adverb*

Either *the* speeches were too long **or** *too* short.

PARALLEL

adverbs

The speeches were **either** *too* long **or** *too* short.

PARALLEL

Either the speeches were too long, **or** they were too short.

18c Use parallel constructions with *and who* and with *and which.*

Avoid *and who, and which,* or *and that* unless they are preceded by a matching *who, which,* or *that.*

NOT PARALLEL
> The position calls for a person with an open mind *and who* is cool-headed.

PARALLEL
> The position calls for a person *who* is open-minded *and who* is cool-headed.

PARALLEL
> The position calls for a person with an open mind and a cool head.

NOT PARALLEL
> A new dam was built to control floods *and that* would furnish recreation.

PARALLEL
> A new dam was built to control floods and to furnish recreation.

PARALLEL
> A new dam was built *that* would control floods *and that* would furnish recreation.

■ Exercise 10

Revise sentences with faulty parallelism. Write C *by correct sentences.*

1. A linguist is one with an interest in the development of
   with
   language and ~~who has~~ a love of words.

2. From an early age the aspiring astronomer believed that his

                                       else
   mission was to look farther into space than anyone‸had looked

            to make
   before and ~~making~~ new discoveries about distances in the

   universe.

3. The characteristics of a good trial lawyer are shrewdness,

                       boldness.
   alertness, and ~~being bold.~~

                              either
4. At the time it appeared ~~either~~ impossible‸to get around the

   crowd or to go through it.

5. Children generally like bubble gum because it is appealingly

                 long lasting.
   packaged, sweet, and ~~it lasts a long time.~~

6. The archaeologists decided to move to another site after

                                 finding
   spending three months digging, sifting, and ~~unable to find~~

   nothing
   ~~anything~~ of significance.

7. A young ballerina must practice long hours, give up pleasures,

   and ~~one has to~~ be able to take severe criticism.

                                performance
8. The ~~performance was~~ long, boring,‸and made everyone in the

   audience restless.

C   9. Roaming through the great north woods, camping by a lake, and getting away from crowds are good ways to forget the cares of civilization.

10. A good listener must have a genuine interest in people, a strong curiosity, and discipline ~~oneself~~ to keep the mind from wandering.

enough (inserted above "discipline oneself")

19 Variety   *var*

Vary sentences in structure and order.

An unbroken series of short sentences may become monotonous and fail to indicate relationships such as cause, condition, concession, time sequence, and purpose (see **11**).

The following discussion of the brain is monotonous because it lacks variety in sentence structure.

> The problem of understanding the brain is a little like that of understanding proteins. Millions of these ingeniously complicated molecular inventions are in every organism. One protein is quite different from the next. It takes years to work out the structural details of even one of them. It takes even longer to know how it works. The prospects are not good for understanding the brain. We do not really understand proteins yet because we do not know how they work. The structure of the brain is somewhat analogous to that of proteins. The brain consists of very large numbers of subdivisions (although not millions). Each has a special architecture and circuit diagram. One is not representative of them all. Hence understanding will be slow. It will come slowly because of practical reasons and

manpower shortage. Progress will be steady (one hopes). It will be asymptotic. There will be breakthroughs but with no likely point of terminus.

Study the variety of the sentences in the passage as it was originally written by David H. Hubel:

> The problem of understanding the brain is a little like that of understanding proteins. There must be millions of those ingeniously complicated molecular inventions in every organism, one protein quite different from the next. To work out the structural details of even one of them seems to take years, to say nothing of knowing exactly how it works. If understanding proteins means knowing how all of them work, the prospects are perhaps not good. In an analogous way the brain consists of very large numbers of subdivisions (although not millions), each with a special architecture and circuit diagram; to describe one is certainly not to describe them all. Hence understanding will be slow (if only for practical reasons of time and manpower), steady (one hopes) and asymptotic, certainly with breakthroughs but with no likely point of terminus.
>
> DAVID H. HUBEL
> "The Brain"

Structure

Do not overuse one kind of sentence structure. Write simple, compound, and complex patterns. Vary your sentences among loose, periodic, and balanced forms.

A **loose sentence,** the most common kind, makes a main point early and then adds further comments and details.

LOOSE

> There are only two or three human stories, and they go on repeating themselves as fiercely as if they had never happened before.
>
> WILLA CATHER

> *Uncle Tom's Cabin* is a very bad novel, having, in its self-righteous, virtuous sentimentality, much in common with *Little Women*.
>
> JAMES BALDWIN

A **periodic sentence** withholds an element of the main thought until the end to create suspense and emphasis.

PERIODIC

> Under a government which imprisons any unjustly, the true place for a just man is also a prison.
>
> HENRY DAVID THOREAU

PERIODIC

> There is one thing above all others that the scientist has a duty to teach to the public and to governments: it is the duty of heresy.
>
> J. BRONOWSKI

A **balanced sentence** has parallel parts that are similar in structure, length, and thoughts. Indeed, *balance* is simply a word for a kind of parallelism (see **18**). The following sentence has perfect symmetry:

> Marriage has many pains, but celibacy has no pleasures.
>
> SAMUEL JOHNSON

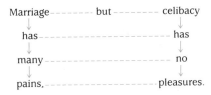

A sentence can also be balanced if only parts of it are symmetrical. The balance in the next sentence consists of nouns in one group that are parallel to adjectives in the other.

Thus the Puritan was made up of two different men, the one all self-abasement, penitence, gratitude, passion; the other proud, calm, inflexible, sagacious.

　　　　　　　　　　　　　　　　THOMAS BABINGTON MACAULAY

Thus
the Puritan
was made up
of two different men,

the one — — — — — —	the other
all self-abasement, — — — —	proud,
penitence, — — — — — —	calm,
gratitude, — — — — — —	inflexible,
passion; — — — — — —	sagacious.

Order

If several consecutive sentences follow the order of subject-verb-complement, they can be monotonous. Invert the order occasionally; do not always tack on dependent clauses and long phrases at the end. Study the following variations:

NORMAL ORDER

subject　　verb　　　　object　　　modifiers ⟶

She attributed these *defects* in her son's character to the general weaknesses of mankind.

SENTENCE BEGINNING WITH DIRECT OBJECT

These *defects* in her son's character she attributed to the general weaknesses of mankind.

SENTENCE BEGINNING WITH PREPOSITIONAL PHRASE

To the general weaknesses of mankind she attributed the defects in her son's character.

SENTENCE BEGINNING WITH ADVERB
> *Quickly* the swordfish broke the surface of the water.

INVERTED SENTENCE BEGINNING WITH CLAUSE USED AS OBJECT
> *That the engineer tried to stop the train,* none would deny.

SENTENCE BEGINNING WITH PARTICIPIAL PHRASE
> *Flying low over the water,* the plane searched for the reef.

■ Exercise 11

Rewrite the following sentences and make them periodic. If you consider a sentence already periodic, write P *by it.*

1. A form of medical treatment from the Orient₁ ~~acupuncture₁~~
: **acupuncture.**
has become popular in America₁∧

P 2. The story that Sir Isaac Newton formulated the law of gravity

after watching an apple fall can be traced to one man, Voltaire.

                                                                   N
3. ~~A sense of humor is one quality that~~ ɴo great leader can be
one quality: a sense of humor.
without₁∧

                        W
4. ~~He studied~~ ʷhen all other possible methods of passing the
, **he studied.**
course proved unworkable₁∧

5. One machine₁ ~~the computer₁~~ is revolutionizing science, stream-

lining business practice, and influencing profoundly the writ-
: **the computer.**
ing habits of many authors₁ ∧

■ Exercise 12

Rewrite the following sentences to give them balanced construc-
tions. Place a B by any sentence that is already balanced.

acting in accordance with that statement
1. Stating that one wishes well for humankind is easy; ~~it~~ is much

   more difficult. ~~to act in accord with that statement.~~

2. The highest peak in North America is Mount McKinley; ~~Mount~~
                                           is Mount Everest.
   ~~Everest is~~ the highest peak in the world.

3. A trained ear hears many separate instruments in an orchestra,
   an untutored ear hears only
   but the melody. ~~is often all that is heard by the untutored.~~

B 4. Realists know their limitations; romantics know only what

   they want.

                                                          an
5. A successful advertisement surprises and pleases, but ~~not all~~
   unsuccessful advertisement bores and irritates.
   ~~advertisements are successful because some are merely boring~~

   ~~and irritating.~~

Punctuation

20 Commas ,

Use commas to reflect structure and to clarify the sense of the sentence.

Commas are chiefly used (1) to separate equal elements, such as independent clauses and items in a series, and (2) to set off modifiers or parenthetical words, phrases, and clauses, all of which take a comma both *before* and *after*.

NOT

The new bridge, a triumph of engineering is attracting national attention.

BUT

The new bridge, a triumph of engineering, is attracting national attention.

20a Use a comma to separate independent clauses joined by a coordinating conjunction (see p. 11).

Nice is a word with many meanings, and some of them are opposite to others.

Sherlock Holmes had to be prepared, for Watson was always full of questions.

The comma is sometimes omitted with a coordinating conjunction between the clauses when they are brief and there is no danger of misreading.

The weather cleared and the aircraft departed.

20b Use commas between words, phrases, or clauses in a series.

> The closet contained worn clothes, old shoes, and dusty hats.

The final comma before *and* in a series is sometimes omitted.

> The closet contained worn clothes, old shoes and dusty hats.

But the comma must be used when *and* is omitted.

> The closet contained worn clothes, old shoes, dusty hats.

And it must be used to avoid misreading.

> An old chest in the corner was filled with nails, hammers, a hacksaw and blades, and a brace and bit.

Series of phrases or of dependent or independent clauses are also separated by commas.

PHRASES
> We hunted for the letter in the album, in the old trunks, and even under the rug.

DEPENDENT CLAUSES
> Finally we concluded that the letter had been burned, that someone had taken it, or that it had never been written.

INDEPENDENT CLAUSES
> We left the attic, Father locked the door, and Mother suggested that we never unlock it again.

In a series of independent clauses, the comma is not omitted before the final element.

■ Exercise 1

Insert commas where necessary in the following sentences.

1. Some prominent women authors took masculine pen names in the nineteenth century, for they felt that the public would not read novels written by women.

2. A good speaker should prepare well for a talk, enunciate clearly to be understood, and practice effective timing.

3. The house with the long driveway is not as old as some others on the block, but none of the historians can determine when it was built.

4. The sales manager and the trainee and a secretary visited branch offices in Pittsburgh, in Dallas, and in San Diego.

5. Rest is often necessary, for the body needs time to recuperate.

6. The sensitive child knew that the earth was round, but she thought that she was on the inside of it.

7. The lecturer exclaimed that Byron was exuberant, that Wordsworth was inspired, that Tennyson was gifted with music, and that Sara Teasdale was profound.

8. For breakfast the menu offered only bacon and eggs, toast and jelly, and hot coffee.

9. Careless driving includes speeding, stopping suddenly, turning from the wrong lane, going through red lights, and failing to yield the right of way.

10. Aspiring actors need publicity for their careers, but some of them never seem to get enough of it.

20c Use a comma between coordinate adjectives not joined by *and*. Do not use a comma between cumulative adjectives.

Coordinate adjectives modify the noun independently.

COORDINATE

We entered a forest of tall, slender, straight pines.

Ferocious, alert, loyal dogs were essential to safety in the Middle Ages.

Cumulative adjectives modify the whole cluster of subsequent adjectives and the noun.

CUMULATIVE

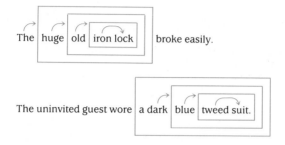

Two tests are helpful.

Test One

And is natural only between coordinate adjectives.

> tall *and* slender *and* straight pines
> ferocious *and* alert *and* loyal dogs

BUT NOT
> dark *and* blue *and* tweed suit
> huge *and* old *and* iron lock

Test Two

Coordinate adjectives are easily reversible.

> straight, slender, tall pines
> loyal, alert, ferocious dogs

BUT NOT
> tweed blue dark suit
> iron old huge lock

The distinction is not always clear-cut, however, and the sense of the cluster must be the deciding factor.

She was wearing a full-skirted, low-cut velvet gown.

[a velvet gown that was full skirted and low cut, not a gown that was full skirted and low cut and velvet]

■ Exercise 2

Insert commas where necessary. When in doubt, apply the tests described above. Write C *by phrases that require no comma.*

   1. a powerful , fearless   hero

C 2. large   glass   front   doors

   3. creamy , soft   vanilla   ice   cream

   4. light , fresh   scent

   5. a   long , shrill , eerie   cry

C 6. a   wrinkled   brown   paper   bag

   7. a   hot , sultry , depressing   day

C 8. expensive   electric   typewriter

   9. a   woebegone , ghostly   look

C 10. shiny   new   automobile

20d   Use a comma after an introductory phrase or clause.

PHRASE

> With the most difficult part of the trek behind him, the traveler felt more confident.

CLAUSE

> When the most difficult part of the trek was behind him, the traveler felt more confident.

NOTE: A comma is always correct after introductory elements, but if the phrase or clause is short and cannot be misread, the comma may be omitted.

SHORT PHRASE

> After the ordeal the traveler felt more confident.

SHORT CLAUSE

> When day came the traveler felt more confident.

BUT

> *comma needed to prevent misreading*
> ↓
> When sixty, people often consider retirement.

Introductory verbal phrases are usually set off by commas.

PARTICIPLE

> Living for centuries, redwoods often reach great heights.

INFINITIVE

> To verify a hypothesis, a scientist performs an experiment.

A phrase or a clause set off by a comma at the beginning of a sentence may not require a comma if it is moved to the end of the sentence (see **21d**).

BEGINNING

> Because they pity creatures that live in cages, some people do not go to the zoo.

END

no comma

> Some people do not go to the zoo because they pity creatures that live in cages.

20e Use commas to set off nonrestrictive appositives, phrases, and clauses.

A **nonrestrictive** modifier or appositive adds information that describes but does not alter the essential meaning of the sentence. When the modifier is omitted, the sentence loses some meaning but does not change radically.

NONRESTRICTIVE

> The painter's latest work, *a landscape*, has achieved wide acclaim.
>
> Salt, *which is plentiful in this country*, is still inexpensive.
>
> Languages contain abstract words, *which do not convey concrete images.*

NOTE: *That* should never introduce a nonrestrictive clause.

NOT

> Salt, *that* is plentiful in this country, is still inexpensive.

BUT

> Salt, *which* is plentiful in this country, is still inexpensive.

A **restrictive** modifier or appositive furnishes information that cannot be removed without radically changing the meaning of the sentence.

RESTRICTIVE

The Russian ruler *Nicholas* was married to Alexandra.

Huge signs *that are displayed along a stretch of Highway 627* advertise a place called The Hornet's Nest.

Words *that convey images* are important in poetry.

In all these sentences, the italicized expressions identify the words they modify; to remove the modifiers would be to change the meaning.

Some modifiers can be either restrictive or nonrestrictive; use or omission of the commas changes the sense.

The coin that gleamed in the sunlight was a Spanish doubloon. [There were several coins.]

The coin, which gleamed in the sunlight, was a Spanish doubloon. [There was only one coin.]

■ Exercise 3

The following pairs of sentences illustrate differences in meaning that result from use or omission of commas with modifiers. Answer the questions about each pair of sentences.

1. A. In Allison Long's novel, *Only Once,* the heroine is a physician.
   B. In Allison Long's novel *Only Once,* the heroine is a physician.

   In which sentence has Allison Long written only one novel? **A**

2. A. People who have faults should not find fault with others.
   B. People, who have faults, should not find fault with others.

   Which sentence implies that all people have faults? **B**

3. A. The winter, which is mild in Virginia, seems to pass rapidly.
   B. The winter that is mild in Virginia seems to pass rapidly.

   Which sentence describes what winters are generally like in Virginia? **A**

4. A. The moviegoers who did not understand Japanese welcomed the English subtitles.
   B. The moviegoers, who did not understand Japanese, welcomed the English subtitles.

   Which sentence suggests that some moviegoers did not depend on English subtitles? **A**

5. A. Anthropologists, who respect native ways, are welcome among most tribes.
   B. Anthropologists who respect native ways are welcome among most tribes.

   Which sentence reflects confidence in anthropologists? **A**

■ Exercise 4

Insert commas for nonrestrictive modifiers; circle all unnecessary commas. Write C by correct sentences.

C 1. The substance known as nicotine was named after Jean Nicot,

   a Frenchman and a lexicographer, who brought the first

   tobacco seedlings to Europe.

2. Barbers⊙who are bald⊙often authoritatively discuss cures for

   balding with their customers⊙who are worried about losing

   their hair.

3. The American composer, George Gershwin, wrote *Rhapsody in Blue,* which was performed last night, on the composer's birthday.

C 4. Americans who have grown up on the prairies may feel shut in when they move to forest regions or to cities with buildings that have more than ten stories.

C 5. According to the Bible, Adam's son Abel was a shepherd.

6. According to the Bible, Abel, Adam's son, was a shepherd.

7. An American flag, that had flown over the Capitol in Washington, was presented to the students by their congresswoman, who needed favorable publicity in her home district.

8. During World War II, American citizens, who were not in jobs essential to the war effort, were issued coupons, with which they could purchase only limited amounts of gasoline.

9. The most beautiful photograph, a shadowy shot of a white bird against a dark sky, was made on a small island that is about a mile off the eastern coast.

10. The character⌒Dorothea Brooke⌒plays an important part in

the novel⌒*Middlemarch*ʌwhich was written by George Eliot,

whose real name was Mary Anne Evans.

20f Use commas to set off sentence modifiers, conjunctive adverbs, and sentence elements out of normal word order.

Modifiers like *on the other hand, for example, in fact, in the first place, I believe, in his opinion, unfortunately,* and *certainly* are set off by commas.

> Some poets, fortunately, do make a living by writing.
> Thomas Hardy's poems, I believe, raise probing questions.

Commas are frequently used with conjunctive adverbs such as *accordingly, anyhow, besides, consequently, furthermore, hence, however, indeed, instead, likewise, meanwhile, moreover, nevertheless, otherwise, still, then, therefore, thus.*

BEFORE CLAUSE
> The auditor checked the figures again; therefore, the mistake was discovered.

WITHIN CLAUSE
> The auditor checked the figures again; the mistake, therefore, was discovered.

Commas **always** separate the conjunctive adverb *however* from the rest of the sentence.

> The auditor found the error in the figures; however, the books still did not balance.
> The auditor found the error in the figures; the books, however, still did not balance.

Commas are not used when *however* is an adverb meaning "no matter how."

> However fast the hare ran, it could not catch the tortoise.

Use commas if necessary for clarity or emphasis when part of a sentence is out of normal order.

> Confident and informed, the young woman invested her own money.

OR

> The young woman, confident and informed, invested her own money.

BUT

> The confident and informed young woman invested her own money.

20g Use commas with degrees and titles and with elements in dates, places, and addresses.

DEGREES AND TITLES

> Sharon Weiss, M.A., applied for the position.
> Louis Ferranti, Jr., owns the tallest building in the city.
> Alphonse Jefferson, chief of police, made the arrest.

DATES

> Sunday, May 31, is her birthday.
> August 1982 was very warm. [Commas around 1982 are also acceptable.]
> July 20, 1969, was the date when a human being first stepped on the moon. [Use commas *before* and *after*.]
> He was born 31 December 1970. [Use no commas.]
> The year 1980 was a time of change. [Restrictive; use no commas.]

PLACES

> Cairo, Illinois, is my home town. [Use commas *before* and *after*.]

ADDRESSES

Write the editor of *The Atlantic*, 8 Arlington Street, Boston, Massachusetts 02116. [Use no comma before the zip code.]

20h Use commas for contrast or emphasis and with short interrogative elements.

The pilot used an auxiliary landing field, not the city airport.
The field was safe enough, wasn't it?

20i Use commas with mild interjections and with words like *yes* and *no*.

Well, no one thought it was possible.
No, it proved to be simple.

20j Use commas with words in direct address and after the salutation of a personal letter.

Driver, stop the bus.
Dear John,
    It has been some time since I've written. . . .

20k Use commas with expressions like *he said, she remarked,* and *she replied* when used with quoted matter.

"I am planning to enroll in Latin," she said, "at the beginning of next term."
He replied, "It's all Greek to me."

20L Set off an absolute phrase with commas.

An **absolute phrase** consists of a noun followed by a modifier. It modifies the sentence as a whole, not any single element in it.

┌──absolute phrase──┐
Our day's journey over, we made camp for the night.

┌──absolute phrase──┐
The portrait having dried, the artist hung it on the wall.

20m Use commas to prevent misreading or to mark an omission.

> After washing and grooming, the poodle looked like a different dog.
>
> When violently angry, elephants trumpet.
>
> Beyond, the open fields sloped gently to the sea.

> *verb omitted*
> ↓
> To err *is* human; to forgive, divine. [Note that *is*, the verb omitted, is the same as the one stated.]

■ Exercise 5

Add necessary commas. If a sentence is correct as it stands, write C by it.

1. Geelong, a city with an unusual name, is in South Victoria, Australia.

2. A few hours before he was scheduled to leave, the mercenary visited his father, who pleaded with him to change his mind and then finally said quietly, "Good luck."

3. Seeing a nightingale, the American ornithologist recognized its resemblance to other members of the thrush family.

4. Inside, the automobile looked almost new; however, it was battered on the outside.

5. History, one would think, ought to teach people not to make the same mistakes again.

6. The Vandyke beard, according to authorities, was named after Sir Anthony Van Dyck, a famous Flemish painter.

C 7. Measles is an acute, infectious viral disease that is recognizable by red circular spots.

8. Having animals for pets is troublesome; having no pets, sometimes lonely.

9. Moving holidays from the middle of the week to Monday, a recent national practice, gives workers more consecutive days without work.

10. While burning, cedar has a distinct, strong odor.

■ Exercise 6

Follow the instructions for Exercise 5.

1. A small and affectionate dog of German breed, the dachshund has a long body, short legs, and drooping ears.

2. Professor Cho does become highly excited sometimes; he is usually, however, calm.

3. A woman of spotless reputation and widely admired wisdom, Abigail Lindstrom, R.N., will be greatly missed.

C 4. Most visitors who see the White House for the first time are surprised because it seems smaller than expected.

5. Found in 1799, the Rosetta stone, a black basalt stone, offered the first clue, fortunately, to the deciphering of Egyptian hieroglyphics.

6. While the mystery writer was composing his last novel, *The Tiger's Eye,* he received a note warning him not to write about anyone he knew in the Orient.

7. The race being over, the jockey who rode the winning horse turned to the owner and said, "Now, Mrs. Astor, you have the money and the trophy."

8. Old encyclopedias, back issues of *National Geographic*, and outdated textbooks were all that remained at the close of the book sale, an eagerly awaited annual event.

9. Highway 280, known as "the world's most beautiful freeway," connects San Francisco and San Jose.

10. Atlanta, Georgia, is lower in latitude than Rome, Italy. Miami, furthermore, is not as far south as the equatorial zone, is it?

■ Exercise 7

Add necessary commas.

1. Bottled water has become popular in many parts of the country, but recent studies show that it may not be any safer than tap water.

2. In J. M. Barrie's play *Peter Pan*, Never-Never-Land is a place of make-believe, a fairyland.

3. Attempting to save money as well as time, some shoppers go to the grocery store only once a month; others, however, go almost daily.

4. The guest fell into a chair, propped his feet on an ottoman, placed his hands behind his head, and yawned as the host glared at him.

5. However the travelers followed the worn, outdated city map, they always returned to the same place.

6. The last selection on the program, a waltz by Strauss, brought the most applause, I believe.

7. High over the mountain, clouds looked dark, ominous.

8. Ruth Friar, Ph.D., was awarded her honorary degree on June 1, 1947, in Fulton, Missouri.

9. Yes, friends, the time has come for pausing, not planning.

10. With a major snowstorm on the way, people should stock up on bread, milk, and eggs, shouldn't they?

21   Unnecessary Commas   *no* **,**

Do not insert commas where they do not belong.

Placing commas at all pauses in sentences is not a correct practice.

21a Do not use a comma to separate subject and verb, verb or verbal and complement, or an adjective and the word it modifies.

NOT

The guard with the drooping mustache, snapped to attention.

Some students in the class admitted, that they had not read, "Mending Wall."

The stubborn, mischievous, child refused to respond.

Two commas may be used to set off a phrase or a word between subject and verb.

CORRECT

The malamute, an Alaskan work dog, can survive extraordinarily cold weather.

21b Do not use a comma to separate two compound elements, such as verbs, subjects, complements, or predicates.

compound verb

He *left* after the first act of the play and *tried* to forget what he had seen.

no comma

21c Do not use a comma before a coordinating conjunction joining two dependent clauses.

dependent clauses

The contractor asserted *that the house was completed* and *that the work had been done properly*.

no comma

21d Do not use a comma before the subordinating conjunction *(after, although, because, before, if, since, unless, until, when, where)* when an adverbial clause follows an independent clause (see **20a** and **20d**).

> *no comma*
> ↓
> We cannot leave today because the storm still rages.

> *no comma*
> ↓
> Remember that it is not necessary to call unless an emergency occurs.

> *no comma*
> ↓
> Ample time will remain in which to ask questions when the lecture ends.

NOTE: In sentences in which the adverbial clause at the end is clearly nonrestrictive, it is usually set off with a comma.

> Joseph Bonaparte, elder brother of Napoleon, reigned as King of Naples until 1808, when he became King of Spain.

21e Do not use a comma before *than* in a comparison.

> *no comma*
> ↓
> The crow is a larger bird than either the cardinal or the Baltimore oriole.

21f Do not use a comma after *like* and *such as.*

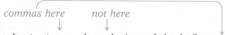

> *commas here*   *not here*
> ↓   ↓
> Some languages, such as Latin and Anglo-Saxon, are no longer spoken.

A comma is not used **before** *such as* (as above) when the phrase is restrictive.

no comma
↓

It was a sight such as she had seen only once or twice before.

21g Do not use a comma with a period, a question mark, an exclamation point, or a dash. These marks stand by themselves.

no comma
↓

"Is the road paved?" asked the tourist.

21h Do not use a comma before an opening parenthesis.

no comma   *comma*
↓   ↓

After a time of inactivity and boredom (nearly six months), Odessa went back to the job from which she had resigned.

21i Do not use a comma to set off coordinating conjunctions.

no comma
↓

And the winner will be announced on Wednesday.

no comma
↓

But that is another story.

21j Do not use commas to set off restrictive clauses, phrases, or appositives (see **20e**).

—— no commas ——

People who live in glass houses should not throw stones.

21k Do not use a comma between adjectives that are not coordinate (see **20c**).

no commas

A dark green canvas awning gave the cottage a look of comfort.

■ Exercise 8

Circle all unnecessary commas; be prepared to explain your decisions.

1. On the flag of the Soviet Union, the hammer and the sickle represent the dignity, and nobility of labor.

2. Tasmanian wolves, were once common, in Australia, but now they are confined to remote parts of Tasmania.

3. Restaurants, that serve excellent food at modest prices, are always popular among local people, though tourists seldom know about them.

4. In the wintry far north of Scandinavia, means of transportation such as snowshoes, skis, and sleds have to be used because vehicles with wheels are impractical.

5. Ticket holders standing at the end of the line worried that they would not find seats or that the only available seats would be too far forward in the theater.

6. Although the composer had indicated that the piece was to be played adagio (slowly), the conductor, sensing the audience's boredom, increased the tempo.

7. Communities near large airports have become increasingly aware that noise pollution can be just as unpleasant as impurities in the air or in streams.

8. Once, huge movie houses were fashionable, but now most of these palaces are like dinosaurs, that is, extinct giants that are curious reminders of the past.

9. The Olympic runner was disqualified after he ran out of his lane, but he would not have won a gold medal anyway.

10. The accountant vowed that she would never work for the millionaire again⌢and that she would go back to her small firm.

■ Exercise 9

Circle all unnecessary commas. Write C *by any correct sentence.*

1. Helen of Troy was⌢the beautiful wife⌢of Menelaus, and her abduction brought on the Trojan War.

2. But⌢the mother of Oedipus was Jocasta, Queen of Thebes.

3. The most famous of all labyrinths⌢was that built by⌢Daedalus⌢for King Minos of Crete.

4. A filmy⌢cobweb wafted on the air⌢until a brisk wind carried it away.

C 5. The prairie dog is a small, quick rodent with a barking cry.

6. Marie Hautbois, who had just joined the orchestra, discovered in an antique shop⌢an oboe⌢that was made in the seventeenth century.

7. She purchased the instrument⌢and told her conductor, Augustine Sey, of her great⌢fortune.

8. Upon first gazing at the Great Pyramid of Khufu, the tourist said that it was "prodigious," and that it was "inspiring beyond measure."

9. He found it grander than the falls that he had visited on the Zambezi River in Africa or the Grand Canyon (which he had seen the previous year), a popular tourist attraction in the United States.

10. Characters with unusual names like Trucker and Egypt are popular among audiences of daytime television drama.

22 Semicolons ;

Use a semicolon between independent clauses not joined by coordinating conjunctions *(and, but, for, nor, or, so, yet)* and between coordinate elements with internal commas.

Omitting a semicolon between independent clauses may result in a comma splice or a fused sentence (see 2).

22a Use a semicolon between independent clauses not connected by a coordinating conjunction.

WITH NO CONNECTIVE

> For fifteen years the painting was stored in the attic; even the artist forgot about it.

WITH A CONJUNCTIVE ADVERB

A specialist from the museum arrived and asked to examine it; *then* all the family became excited.

See **20f** for use of commas with conjunctive adverbs.

WITH A SENTENCE MODIFIER

The painting was valuable; *in fact,* the museum offered ten thousand dollars for it.

See **20f** for use of commas with sentence modifiers, such as *on the other hand, for example, in fact, in the first place.*

Notice the use of semicolons in the paragraph below, a light-hearted treatment of semicolons.

> The semicolon tells you that there is still some question about the preceding full sentence; something needs to be added; it reminds you sometimes of the Greek usage. It is almost always a greater pleasure to come across a semicolon than a period. The period tells you that that is that; if you did not get all the meaning you wanted or expected, anyway you got all the writer intended to parcel out and now you have to move along. But with a semicolon there you get a pleasant little feeling of expectancy; there is more to come; read on; it will get clearer.
>
> LEWIS THOMAS
> *The Medusa and the Snail*

22b   Use a semicolon to separate independent clauses that are long and complex or that have internal punctuation.

In many compound sentences either a semicolon or a comma can be used.

COMMA OR SEMICOLON

Moby-Dick, by Melville, is an adventure story, [*or* ;] and it is also one of the world's greatest philosophical novels.

SEMICOLON PREFERRED

> Ishmael, the narrator, goes to sea, he says, "whenever it is a damp, drizzly November" in his soul; and Ahab, the captain of the ship, goes to sea because of his obsession to hunt and kill the great white whale, Moby Dick.

22c Use semicolons in a series between items that have internal punctuation.

> The small reference library included a few current periodicals, those most often read; a set of encyclopedias (The *Americana,* I believe); several dictionaries, both abridged and unabridged; and various bibliographical tools.

22d Do not use a semicolon between elements that are not coordinate.

FAULTY

> ┌────────────── *dependent clause* ──────────────┐   *independent*
> After the tugboat had signaled to the barge twice; it turned toward
>   *clause*   ↑
> the wharf.   *use* ,

■ Exercise 10

Circle unnecessary semicolons and commas, and insert necessary ones. Write C by sentences that are correct.

1. Isobars are, lines on maps, that connect points having the same barometric pressure.

2. The assignment was, to seek out a successful executive and to interview him or her, asking particular questions about educational training, and experience.

C   3. An advanced civilization is guided by enlightened self-interest; however, it is also marked by unselfish good will.

     4. The Pawnees, who lived at one time in the valley of the Platte River in Nebraska, were once a numerous people **;** but their numbers diminished when the group, part of a confederacy of North American Plains Indians, moved from their ancestral homes **,** to northern Oklahoma.

     5. Because Irving was careful **,** his friends accused him of being indecisive.

     6. The hallway was long **,** and dark, *or ;* and at the end of it hung a dim, obscure painting representing a beggar **,** in eighteenth-century London.

     7. The soybean plant is now widely cultivated in America **;** it is native, however, to China and Japan.

     8. The storm knocked down the power lines, leaving the town dark **;** uprooted trees, leaving the streets blocked **;** and forced water over the levee, leaving the neighborhoods near the river flooded.

9. Fortunetelling still appeals to many people; even when they realize it is superstitious nonsense, they continue to patronize charlatans like palm readers.

10. The making of pottery, once a necessary craft as well as an art, has again become popular; and hundreds of young people, many of them highly skillful, have discovered the excitement of this art.

23 Colons :

Use a colon as a formal mark of introduction.

23a Use a colon after an independent clause that introduces a quotation or a series of items (see 23e).

BEFORE A QUOTATION
On the locket was an engraved reminder: "Forever yours."

BEFORE A SERIES
An excellent physician exhibits four broad characteristics: knowledge, skill, compassion, and integrity.

23b Use a colon after an independent clause that introduces an appositive.

An **appositive** is a word, phrase, or clause used as a noun and placed beside another word to explain, identify, or rename it.

> One factor is often missing from modern labor: pleasure in work. [Pleasure in work *is* the factor.]
>
> The author made a difficult decision: he would abandon the script. [To abandon the script *is* the decision.]

Frequently appositives are preceded by expressions like *namely* and *that is*.

> The new law calls for a big change: namely, truth in advertising. [Note that the colon comes **before** *namely*, not after.]

23c Use a colon between two independent clauses when one explains the other.

> Music communicates: it is an expression of deep feeling.

23d Use a colon after the salutation of a formal letter, between figures indicating hours and minutes, and in bibliographical entries.

> Dear Dr. Tyndale: *PMLA* 99 (1984): 75
> 12:15 P.M.   Boston: Houghton, 1929

23e Do not use a colon after a linking verb or after a preposition.

NOT AFTER LINKING VERB

$$no\ colon$$
$$\downarrow$$

Some chief noisemakers **are** automobiles and airplanes.

NOT AFTER PREPOSITION

no colon
↓

His friend accused him **of** wiggling in his seat, talking during the lecture, and not remembering what was said.

24 Dashes —

Use a dash to introduce summaries and to indicate interruptions, parenthetical remarks, and special emphasis.

NOTE: In typing, a dash is made by two hyphens (--) with no space before or after it.

FOR SUMMARY

Attic fans, window fans, air conditioners——all were ineffective that summer.

FOR SUDDEN INTERRUPTIONS

She replied, "I will consider the——No, I won't either."

FOR PARENTHETICAL REMARKS

A true poet in moments of greatest inspiration——this was Robbie Ridley's argument——is not conscious of being a poet.

FOR SPECIAL EMPHASIS

Great authors quote one book more than any other——the Bible.

25 Parentheses ( )

Use parentheses to enclose a loosely related comment or explanation, figures that number items in a series, and references in documentation.

FOR A COMMENT

The frisky colt (it was not a thoroughbred) brought a good price at the auction.

A parenthetical sentence within another sentence has no period or capital, as in the example above. A freestanding parenthetical sentence requires parentheses, a capital, and a period.

capital ↓ *period here* ↓

On that day all flights were on time. (The weather was clear.)

FOR FIGURES

The investor refused to buy the land because (1) it was too remote, (2) it was too expensive, and (3) the owner did not have a clear title.

FOR A REFERENCE IN DOCUMENTATION

Link does not agree (432). [See pp. 367–368, 426.]

26 Brackets [ ]

Use brackets to enclose interpolations within quotations.

In the opinion of Arthur Miller, "There is no more reason for falling down in a faint before his [Aristotle's] *Poetics* than before Euclid's geometry."

Parenthetical elements within parentheses are indicated by brackets ([ ]). Try to avoid constructions that call for this intricate punctuation.

27 Quotation Marks " "

Use quotation marks to enclose the exact words of a speaker or writer and to set off some titles.

Most American writers and publishers use double quotation marks (". . .") except for internal quotations, which are set off by single quotation marks ('. . .') (see **27b**).

27a Use quotation marks to enclose direct quotations and dialogue.

DIRECT QUOTATION

At a high point in *King Lear,* the Duke of Gloucester says, "As flies to wanton boys are we to the gods."

NOTE: Do *not* use quotation marks to enclose indirect quotations.

He said that the gods regard us as flies.

In dialogue a new paragraph signals each change of speaker.

DIALOGUE

"What is fool's gold?" asked the traveler who had never before been prospecting.

"Well," the geologist told him, "it's pyrite."

NOTE: Do not use quotation marks to enclose prose quotations that are longer than four lines. Instead, indicate the quotation by blocking—indenting ten spaces from your left margin and single-spacing or double-spacing according to the preference of your instructor.

Unless your instructor specifies otherwise, poetry of four lines or more should be double-spaced and indented ten spaces. Retain the original divisions of the lines.

```
If you would keep your soul

From spotted sight or sound,

Live like the velvet mole;

Go burrow underground.
```

Quotations of three lines of poetry or less may be written like the regular text—not set off. Use a slash (with a space before and after) to separate lines:

Elinor Wylie satirically advises, "Live like the velvet mole; / Go burrow underground."

27b Use single quotation marks to enclose a quotation within a quotation.

The review explained: "Elinor Wylie is ironic when she advises, 'Go burrow underground.'"

27c Use quotation marks to enclose the titles of essays, articles, short stories, short poems, chapters (and other subdivisions of books or periodicals), dissertations (see pp. 361–367), episodes of television programs, and short musical compositions.

D. H. Lawrence's "The Rocking-Horse Winner" is a story about the need for love.

One chapter of *Walden* is entitled "The Beanfield." [For titles of books, see **30a**.]

27d   On your paper, do not use quotation marks around its title.

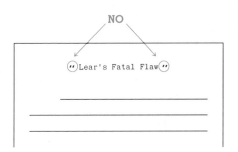

NO

"Lear's Fatal Flaw"

27e   Do not use quotation marks to emphasize or change the usual meanings of words or to justify slang, irony, or attempts at humor.

no

The beggar considered himself a "rich" man.

no

The old politician's enemies hoped that he would "croak."

Quotation marks do not give specialized or unusual definitions to words. They do not effectively add new meanings.

27f   Do not enclose a block (set-off) quotation in quotation marks (see **27a** and pp. 382–383).

Do not add quotation marks to show that you are quoting the entire passage. Blocking does that. Within the block passage use quotation marks exactly as they appear in the material being quoted.

27g Follow established conventions in placing other marks of punctuation inside or outside closing quotation marks.

Periods and **commas** in American usage are placed *inside* closing quotation marks.

> All the students had read "Lycidas."
> "Amazing," the professor remarked.

Semicolons and **colons** are placed *outside* closing quotation marks.

> The customer wrote that she was "not yet ready to buy the first edition"; it was too expensive.

A **question mark** or an **exclamation point** is placed *inside* closing quotation marks when the quotation itself is a direct question or an exclamation. Otherwise, these marks are placed *outside*.

> He asked, "Who is she?" [Only the quotation is a question.]
> "Who is she?" he asked. [Only the quotation is a question.]
> Did he ask, "Who is she?" [A quoted question within a question takes only one question mark—inside the quotation marks.]
> Did he say, "I know her"? [The entire sentence asks a question; the quotation makes a statement.]
> She screamed, "Run!" [Only the quotation is an exclamation.]
> Curse the man who whispers, "No"! [The entire statement is an exclamation; the quotation is not.]

After quotations, do not use a period or a comma together with an exclamation point or a question mark.

> NOT
> "When?", I asked.
>
> BUT
> "When?" I asked.

For parentheses with quotations and page numbers in documentation, see pp. 367–368, 426.

28 End Punctuation .?!

Use periods, question marks, or exclamation points to end sentences and to serve special functions.

28a Use a period after a sentence that makes a statement or expresses a command.

Some modern people claim to practice witchcraft.

Water the flowers.

The gardener asked whether the plant should be taken indoors. [This sentence is a statement even though it expresses an indirect question.]

28b Use periods after most abbreviations.

Periods follow such abbreviations as Mr., Dr., Pvt., Ave., B.C., A.M., Ph.D., e.g., and many others. In British usage, periods are often omitted after titles (Mr).

Abbreviations of governmental and international agencies often are written without periods (FCC, TVA, UNESCO, NATO, and so forth). Usage varies. Consult your dictionary.

A comma or another mark of punctuation may follow the period after an abbreviation, but at the end of a sentence only one period is used.

After she earned her M.A., she began studying for her Ph.D.

But if the sentence is a question or an exclamation, the end punctuation mark follows the period after the abbreviation.

When does she expect to get her Ph.D.?

28c Use three spaced periods (ellipsis points) to show an omission in a quotation.

Note how the following quotation can be shortened with ellipsis points:

> The drab, severe costumes of the Puritan settlers of New England, and their suspicion of color and ornaments as snares of the devil, have left their mark on the present-day clothes of New Englanders. At any large meeting, people from this part of the country will be dressed in darker hues—notably black, gray, and navy—often with touches of white that recall the starched collars and cuffs of Puritan costume. Fabrics will be plainer (though heavier and sometimes more expensive) and styles simpler, with less waste of material. Skirts and lapels and trimmings will be narrower.

> ALISON LURIE
> *The Language of Clothes*

ELLIPSIS POINTS TO INDICATE OMISSIONS

ellipsis points not necessary here
↓

Clothing "of the Puritan settlers of New England, and their suspicion of color and ornament . . . , have left their mark on the present-day clothes of New Englanders. At any large meeting, people from this part of the country will be dressed in darker hues . . . often with touches of white that recall the starched collars and cuffs of Puritan costume. Fabrics will be plainer . . . and styles simpler. . . ."

Use three ellipsis points for omissions within sentences.
Use period plus three ellipsis points for omission at end of sentence.

NOTE: Although the comma after ellipsis points in the first sentence above could be omitted, it is permissible to include it since it completes a parenthetical element of the sentence.

28d A title does not end with a period even when it is a complete sentence, but some titles include a question mark or an exclamation point.

Nobody Knows My Name   "What Are Years?"
Absalom, Absalom!   *Ah! Wilderness*

28e Use a question mark after a direct question.

Do teachers file attendance reports?
Teachers do file attendance reports? [a question in declarative form]

Question marks may follow separate questions within an interrogative sentence.

Do you recall the time of the accident? the license numbers of the cars involved? the names of the drivers?

NOTE: A question mark within parentheses shows that a date or a figure is historically doubtful.

Pythagoras, who died in 497 B.C. (?), was a philosopher.

28f Do not use a parenthetical question mark or exclamation point to indicate humor or sarcasm.

NOT
The comedy (?) was a miserable failure.

28g   Use an exclamation point after a word, a phrase, or a sentence to indicate strong exclamatory feeling.

> Wait! I forgot my lunch!
> What a ridiculous idea!
> Stop the bus!

Use exclamation points sparingly. After mild exclamations, use commas or periods.

NOT
> Well! I was discouraged!

BUT
> Well, I was discouraged.

■ Exercise 11

Supply quotation marks as needed in the following passage, and insert the sign ¶ where new paragraphs are necessary.

There was a pause. Then I knocked again. And almost immediately a light sprang up above and an upper window opened. ¶"Wer ist da?" cried a man's voice. ¶ I felt the shock of disappointment and consternation to my fingers. ¶"I want help; I have had an accident," I replied. ¶ Some muttering followed. Then I heard steps descending the stairs, the bolt of the door was drawn, the lock was turned. It was opened abruptly, and in the darkness of the passage a tall man hastily attired, with a pale face and dark moustache, stood before me. "What do you want? he said, this time in English. I had now to think of something to say. I wanted above all to get into parley with this man, to get matters in such a state that instead of raising an alarm and summoning others he would discuss things quietly. "I am a burgher, I began. "I have had an accident. I was going to join my commando at Komati Poort. I have fallen off the train. We were skylarking. I have been unconscious for hours. I think I have dislocated my shoulder." It is astonishing how one thinks of these things. This story leapt out as if I had learnt it by heart. Yet I had not the slightest idea what I was going to say or what the next sentence would be. . . . "I think I'd like to know a little more about this railway accident of yours," he said, after a considerable pause. "I think, I replied, "I had better tell you the truth." "I think you had," he said, slowly. So I took the plunge and threw all I had upon the board. "I am Winston Churchill, War Correspondent of the *Morning Post*. I escaped last night from Pretoria. I am making my way to the frontier. I have plenty of money. Will you help me?[„¶] There was another long pause. My

companion rose from the table slowly and locked the door. After

this act, which struck me as unpromising, and was certainly

ambiguous, he advanced upon me and suddenly held out his

hand.[¶“] Thank God you have come here! It is the only house for

twenty miles where you would not have been handed over. But

we are all British here, and we will see you through.["]

WINSTON S. CHURCHILL
My Early Life: A Roving Commission

■ Exercise 12

*Add quotation marks where needed; circle unnecessary ones. Also
make all necessary changes in punctuation.*

1. "Failure is often necessary for humanity," the speaker said.

   ["]Without failure,["] he continued,["] how can we retain our humility

   and know the full sweetness of success? For, as Emily

   Dickinson said,['] Success is counted sweetest / By those who

   ne'er succeed.[']["]

2. ["]Madam,["] said the talent scout,["] I know that you think your

   daughter can sing, but, believe me, her voice makes the

   strangest sounds I have ever heard.["] Mrs. Audubon took her daughter "Birdie" by the hand and haughtily left the room wondering "how she could ever have been so stupid as to expose her daughter to such a "common" person."

3. Grandmother said, "I'm going to teach you a poem that I learned when I was a little girl. It's called 'The Wind,' and it begins like this: 'I saw you toss the kites on high / And blow the birds about the sky.'"

4. The boy and his great-uncle were "real" friends, and the youngster would listen intently when the old man spoke. "Son," he would say, "I remember my father's words: 'You can't do better than to follow the advice of Ben Franklin, who said, "One To-day is worth two To-morrows."'"

5. To demonstrate that the English language is always "chang-ing" the teacher said that "each student should come up with" a list of expressions, such as "boom box," "courseware," and "golden parachute," not found in the dictionary used in class.

6. A recent report states the following: "The marked increase in common stocks indicated a new sense of national security;" however, the report seems to imply "that this is only one of many gauges of the country's economic situation.

7. Chapters in modern novels rarely have any titles at all, especially "wordy" ones like the title of Chapter 51 in *Vanity Fair* (1848): "In Which a Charade Is Acted Which May or May Not Puzzle the Reader."

8. On a hotel postcard sent to his "former fiancée," the young man marked an "X" on the picture of the building by the room where he was spending his "honeymoon" with his new bride.

9. "This is where we are staying," he wrote. "Wish you were here."

10. It was not at all "unusual" for the wanderer to stop busy passers-by on the street and ask them "why they were hurrying when life is so short."

Mechanics

29   Manuscript and Letter Form   *ms*

Follow correct manuscript form in papers and business letters.

29a   Follow requirements and conventions in preparing a manuscript.

Paper

    typing: white, 8½ by 11 inches
    longhand: ruled
    not: onionskin, spiral, legal size

Lines

    typing: double-spaced
    longhand: skip every other line (or follow your teacher's instructions)

Title

    centered with extra space between title and text

Margins

    ample and regular at bottom and top—at least one inch on each side

Page numbers

    Arabic numerals (2, *not* II) in upper right corner (optional on first page)

Example of correct manuscript form

about 1½" margin

indent 5 spaces

The Recurring Hoax ←——————— *center*

←——————— *triple-space*

Erik Demos has proved that there is indeed a difference between a

2 spaces after periods

good book and a good hoax: the hoax survives longer.↓ Demos is the

pseudonym of freelance writer Chuck Ross, who, in 1975, submitted an

excellent novel entitled Steps to four New York publishers. He knew the

manuscript was a good one because it had won the National Book Award

when it was first published by Random House in 1969. The real author

was Jerzy Kosinski, but it was Demos who retyped the book and submitted

it to prove a point, that a first novelist has virtually no chance to be

published. All four publishers cooperated by rejecting the submission.

A trade biweekly called Bookletter carried the story in 1975 and

mentioned the "author's" intention of trying again with his second-hand

novel. He did, this time in 1977. Steps went to fourteen publishers;

all of them turned it down, the most embarrassed of the lot being Random

House, the original publisher. One or two others mentioned that the

style seemed imitative of Jerzy Kosinski, but the only encouragement

came in the form of suggestions to try literary agents. So Demos tried.

Twelve agents would not bother with the manuscript. The Knox Burger

agency lost it. On February 19, 1979, Time ran a short article on the

hoax and announced that it was about time for it to recur, saying fur-

ther that the conspirators [Ross and Kosinski] were determined to elicit

2 hyphens no space for dash

some form of positive response from the publishing community—either a

contract or recognition of the novel for what it was. In the meantime

Kosinski advises fledgling writers to go into accounting.

about 1½" margin at
bottom of page

RANDY F. NELSON
The Almanac of American Letters

29b   Follow accepted practices in preparing business letters.

In writing a business letter follow conventional forms. All essential parts are included in the examples on pp. 165–167. Type or print out; single-space, with double-spacing between paragraphs. Paragraphs may begin at the left margin without indentation (block form), or they may be indented.

Try to learn the name *and* the title of the addressee. With names, use *Mr.* for men and *Ms.* for women (unless the recipient has indicated a preference for *Mrs.* or *Miss*).

Opinions vary about the best salutation when the name, title, or sex of the person is unknown. If the sex of the addressee is known but the name is not, use *Dear Sir* or *Dear Madam*. If the sex is not known, use *Dear Sir or Madam,* or omit the salutation entirely.

Business letters are usually written on stationery measuring 8½ by 11 inches. Fold horizontally into thirds to fit a standard-sized business envelope. For smaller envelopes fold the letter once horizontally and twice the other way. (See page 167.)

Block form

601 Longview Terrace, Apt.   ⟵——— *address of writer*
Manhattan, KS 66506
March 7, 1991 ⟵——————————— *date*

Ms. Christina T. Arrowsmith ⟵——— *name and title of addressee*
Vice President for Personnel
Lucerne Overton Company
6 Beacon Street
Boston, MA 02108 ⟵——————————— *full address*

Dear Ms. Arrowsmith: ⟵——————— *salutation and name (use a colon)*

I am writing to you at the suggestion of my adviser at Kansas State
University, Professor Leland Levenson, who has strongly urged me to
pursue a career in commercial publishing. For some time I have been
investigating the opportunities and challenges in that field. The more
I learn, the more I am convinced that I wish to be part of a company
with the long history and the distinction of Lucerne Overton.

This term I shall graduate with a B.A. degree and with a joint major in
English and history. My overall grade point average is 3.15 with an
average in my major of 3.56. My training and aptitudes are especially
strong in areas related to publishing. I have taken an undergraduate
seminar in editing; a class in the history of bookmaking; and two
courses in business, one of which dealt with sales. I am enclosing a
résumé, which will give you a more detailed summary of my background.
I wish to apply for any position open at Lucerne Overton for which you
consider me suited.

I shall be grateful for any advice that you can offer or any help you
can give me. If you think it appropriate, I shall be glad to come to
Boston for an interview. It would certainly be my pleasure to meet you
and to have an opportunity to see the home office of a company for which
I have admiration and respect.

Yours truly, ⟵————————————— *complimentary close*

Patrick O'Halihan ⟵————— *signature, handwritten*   ⎫ 3-line
Patrick O'Halihan ⟵————— *name, typed*   ⎬ space
                                                           ⎭

Enclosure ⟵————————————— *indication of material enclosed*

Modified block form

141 Oakhurst Drive, Apt. 2A
Singleton, OH 54567
March 1, 1991

Mr. Freeman O. Zachary, Manager
Personnel Department
Keeson National Bank
P. O. Box 2387
Chicago, IL 34802

Dear Mr. Zachary:

I am writing to ask whether you will have an opening this coming summer
for someone with my qualifications. I am finishing my sophomore year at
Singleton State College, where I intend to major in economics and fi-
nance. I have been active in several extracurricular activities, including the Spanish Club and the Singleton Players.

Although I have no previous experience in banking, I am eager to learn,
and I am willing to take on any duties you feel appropriate. Chicago is
my home town, so I am especially anxious to find summer employment
there. I will be in Chicago for spring vacation from March 20 to 26 and
will be available for an interview. In addition, I shall be pleased to
furnish you with letters of recommendation from some of my professors
here at Singleton State and with a transcript of my college record. I
appreciate your consideration, and I look forward to hearing from you.

Sincerely yours,

Audrey DeVeers

Audrey DeVeers

Audrey DeVeers
141 Oakhurst Drive, Apt. 2A
Singleton, OH 54567

```
┌ ─ ─ ┐
  Place
  stamp
  here
└ ─ ─ ┘
```

Mr. Freeman O. Zachary, Manager
Personnel Department
Keeson National Bank
P. O. Box 2387
Chicago, IL 34802

Indented form

7315 Altamont Circle
Green Bay, WI 54311
July 27, 1991

Mr. Andrew Y. Liephart, President
Leakproof Plumbing and Heating Co.
837 Hasty Industrial Way
Green Bay, WI 54311

Dear Mr. Liephart:

On Friday, July 3, two of your employees installed a new electric
hot-water heater in my residence. In removing the old water heater, the
workers allowed water containing rust to spill out and stain the carpeting in my finished basement. They acknowledged the accident, but
they indicated that such occurrences are common and almost inevitable. I
do not accept this explanation, and I wish to point out that I signed no
statement relieving your company of liability incurred through
damages to my property.

I have had the carpeting professionally cleaned, and I feel that
your company should reimburse me for this expense. I enclose a copy of
the bill from Likenew Carpet Cleaners in the amount of $93.16. I expect
to receive a check from you in this amount so that this matter can be
settled without my having to take further action.

Sincerely yours,

Lucille Navarro

Lucille Navarro

Enclosure

Folding the letter

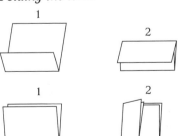

30   Underlining for Italics   *ital*

Underline titles of separate publications (books, magazines, newspapers), and underline *occasionally* for emphasis.

Italic type slants *(like this)*. Underline words separately (like this), or underline words and spaces (like this).

30a   Underline titles of books (except the Bible and other sacred books and their divisions), periodicals, newspapers, motion pictures, paintings, sculptures, musical compositions, television and radio programs, plays, and other works published separately.

Be precise: underline initial articles *(A, An, The)* and any punctuation when they are part of the title.

BOOKS

    Adventures of Huckleberry Finn (*not* The Adventures . . .)

    An American Tragedy (*not* The American Tragedy)

    Not underlined: the Bible, John 3:16, the Koran

PERIODICALS

    The Atlantic and the American Quarterly

NEWSPAPERS

    The New York Times or the New York Times or the New York Times

MOTION PICTURES

    Citizen Kane

MUSICAL COMPOSITIONS
> Bizet's <u>Carmen</u>
> Beethoven's <u>Mount of Olives</u> (*but not* Beethoven's Symphony No. 5)

PLAYS
> <u>The Cherry Orchard</u>

30b Underline names of ships and trains.

> the <u>Queen Elizabeth II</u>   the U.S.S. <u>Hornet</u>   the <u>Zephyr</u>

30c Underline foreign words used in an English context except words that have become part of our language.

Consult a dictionary to determine whether a word is still considered foreign or has become part of the English language.

> He claimed extravagantly to have been <u>au courant</u> since birth.
> The dandelion, of the genus <u>Taraxacum</u>, is a common weed.

BUT
Some words that may seem foreign have become a part of the English language.

> faux pax, amigo, karate

30d Underline words, letters, and figures being named.

> Remember to cross your <u>t</u>'s. [or <u>ts</u>]
> Glee rhymes with <u>flea</u>.

NOTE: Occasionally quotation marks are used instead of underlining.

30e Avoid frequent underlining for emphasis.

Numerous underlinings, dashes, and exclamation points on a page are distracting. Do not attempt to create excitement by overusing these marks.

30f Do not underline the title of your paper.

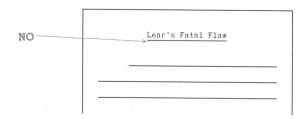

NO ——————→ Lear's Fatal Flaw

■ Exercise 1

Underline words as necessary in the following sentences. Put an X over words unnecessarily underlined. Place a C over words correctly underlined.

      X   X

1. The cotton gin is often associated with the industrial revolu-

   tion.

      C   C

2. Newsweek reviewed the new play Heaven Cannot Wait.

3. The author's last novel, The Green Summer, is based upon a

              X   X   X   X

   story in the Bible and is a fine work, not to be missed.

4. The painting is titled Arrangement in Gray and Black.

        X
5. The limousine was used in a motion picture entitled The

    Years of Hope.

6. The motion picture The Killers is based on a short story by

    Hemingway.

7. Jack London's short story "To Build a Fire" is published in

    the latest edition of the anthology America's Great Tales.

8. The periodical Harper's has an article on the modern opera

    Streets of the City.

                                                          X
9. He made his i's with little circles over them instead of dots.

10. While on the train The Northern Star, she saw a large moose

      C
    (Alces americana).

31 Spelling   *sp*

Spell correctly; make a habit of checking words in your dictionary.

Spelling is troublesome in English because many words are not spelled as they sound *(laughter, slaughter)*; because some pairs and triplets sound the same *(capital, capitol; there, they're, their; to, too, two)*; and because many words are pronounced with the vowel sound "uh," which gives no clue to spelling (sensible, capable, defiant).

Many misspellings result from the omission of syllables in habitual mispronunciations *(accident-ly* for *acciden-tal-ly);* the addition of syllables *(disas-ter-ous* for *disas-trous);* or the changing of syllables *(prespiration* for *perspiration).*

English has no infallible guides to spelling, but the following are helpful.

ie or *ei?*

Use *i* before *e*
Except after *c*
Or when sounded as *a*
As in n*ei*ghbor and w*ei*gh.

WORDS WITH *IE*
bel*ie*ve, ch*ie*f, f*ie*ld, gr*ie*f, p*ie*ce

WORDS WITH *EI* AFTER *C*
rec*ei*ve, rec*ei*pt, c*ei*ling, dec*ei*t, conc*ei*ve

WORDS WITH *EI* SOUNDED AS *A*
fr*ei*ght, v*ei*n, r*ei*gn

EXCEPTIONS TO MEMORIZE
*ei*ther, n*ei*ther, l*ei*sure, s*ei*ze, w*ei*rd, h*ei*ght

Drop final silent e?

DROP		KEEP	
When suffix begins with a vowel		*When suffix begins with a consonant*	
curse	cursing	live	lively
come	coming	nine	ninety
pursue	pursuing	hope	hopeful
arrange	arranging	love	loveless
dine	dining	arrange	arrangement

TYPICAL EXCEPTIONS

courageous
noticeable
dyeing (compare *dying*)
singeing (compare *singing*)

TYPICAL EXCEPTIONS

awful
ninth
truly
argument

Change y to i?

CHANGE
*When y is preceded by
a consonant*

gully	gullies
try	tried
fly	flies
apply	applied
party	parties

DO NOT CHANGE
*When y is preceded by
a vowel*

valley	valleys
attorney	attorneys
convey	conveyed
pay	pays
deploy	deploying

When adding -ing

try	trying
fly	flying
apply	applying

Double final consonant?

If the suffix begins with a consonant, do not double the final consonant of the base word *(man, manly)*.

If the suffix begins with a vowel:

DOUBLE
*When final consonant is
preceded by single vowel*

Monosyllables

pen	penned
blot	blotted
hop	hopper
sit	sitting

DO NOT DOUBLE
*When final consonant is
preceded by two vowels*

despair	despairing
leer	leering

*Words ending with two or more
consonants preceded by single
vowel*

jump	jumping
work	working

DOUBLE	DO NOT DOUBLE

Polysyllables accented on last syllable	*Polysyllables not accented on last syllable after addition of suffix*

defér	defe**rr**ing	defér	déference
begín	begi**nn**ing	prefér	préference
omít	omi**tt**ing	devélop	devéloping
occúr	occu**rr**ing	lábor	lábored

Add s *or* es?

ADD *S*

For plurals of most nouns

girl	girl**s**
book	book**s**

For nouns ending in o *preceded by a vowel*

radio	radio**s**
cameo	cameo**s**

ADD *ES*

When the plural is pronounced as another syllable

church	church**es**
fox	fox**es**

Usually for nouns ending in o *preceded by a consonant (consult your dictionary)*

potato**es**
hero**es**

BUT

flamingos *or* flamingoes
egos

NOTE: The plurals of proper names are generally formed by adding *s* or *es* (*Darby*, the *Darbys*; *Jones*, the *Joneses*).

Words frequently misspelled

Following is a list of over two hundred of the most commonly misspelled words in the English language.

absence	accumulate	advice
accessible	accuracy	advise
accidentally	acquaintance	all right
accommodate	acquitted	altar

amateur
among
analysis
analyze
annual
apartment
apparatus
apparent
appearance
arctic
argument
arithmetic
ascend
athletic
attendance
balance
beginning
believe
benefited
boundaries
Britain
business
calendar
candidate
category
cemetery
changeable
changing
choose
chose
coming
commission
committee
comparative
compelled
conceivable
conferred
congratulations
conscience
conscientious
control
convenient
criticize
deceive

deferred
definite
democracy
description
desperate
dictionary
dining
disappearance
disappoint
disastrous
discipline
dissatisfied
dormitory
ecstatic
eighth
eligible
eliminate
embarrass
eminent
encouraging
environment
equipped
especially
exaggerate
excellence
exhilarate
existence
experience
explanation
familiar
fascinate
February
fiery
foreign
formerly
forty
fourth
frantically
fulfill or fulfil
generally
government
grammar
grandeur
grievous

guaranteed
harass
height
heroes
hindrance
hoping
humorous
hypocrisy
immediately
incidentally
incredible
independence
inevitable
intellectual
intelligence
interesting
irresistible
knowledge
laboratory
laid
led
lightning
loneliness
maintenance
management
maneuver
manual
manufacture
marriage
mathematics
may
maybe
miniature
mischievous
mysterious
necessary
ninety
noticeable
occasionally
occurred
omitted
opportunity
optimistic
parallel

paralyze	prophesy	sincerely
pastime	quantity	sophomore
perceive	quiet	specifically
performance	quite	specimen
permanent	quizzes	stationary
permissible	recede	stationery
perseverance	receive	statue
personnel	recognize	studying
perspiration	recommend	subtly
physical	reference	succeed
picnicking	referred	successful
playwright	repetition	summary
possibility	restaurant	supersede
practically	rhythm	suppose
precede	ridiculous	surprise
precedence	roommate	temperamental
preference	sacrifice	tendency
preferred	salary	their
prejudice	schedule	thorough
preparation	secretary	through
prescribe	seize	vegetable
prevalent	separate	vengeance
privilege	sergeant	villain
probably	severely	weird
professor	shining	writing
pronunciation	siege	yield
prophecy	similar	

■ Exercise 2

In each of the following groups of words one, two, or three are misspelled. The others are correct. Put an X over incorrectly spelled words.

      X   X   X

1. translate, translat, transerlate, persume, presume

   X   X

2. reminent, remnant, perifery, tangent, tangerine

   X   X

3. percieve, believe, recieve, achieve, conceive

      X   X   X
4. mountain, villian, protein, maintainance, certian

             X   X   X
5. credence, precedence, balence, existance, independance

  X   X   X
6. tallys, valleys, bellys, modifys, fancies

  X   X   X
7. defys, relays, conveyes, carries, dirtys

          X   X   X
8. obedience, modifyer, complience, applience, guidance

              X   X
9. incredible, detectable, delectible, dependible, reversible

  X   X   X
10. sensable, receivable, edable, likible, noticeable

32   Hyphenation and Syllabication   -

Use a hyphen in certain compound words and in words divided at the end of a line.

Two words not listed as an entry in a dictionary are usually written separately *(campaign promise)*.

32a   Consult a dictionary to determine whether a compound is hyphenated or written as one or two words (see 37a).

HYPHENATED	ONE WORD	TWO WORDS
drop-off (noun)	droplight	drop leaf (noun)
white-hot	whitewash	white heat
water-cool	watermelon	water system

32b Hyphenate a compound of two or more words used as a single modifier before a noun.

HYPHEN   NO HYPHEN AFTER NOUN
She is a *well-known* executive.   The executive is *well known*.

A hyphen is not used when the first word of such a group is an adverb ending in *-ly*.

HYPHEN   NO HYPHEN
a *half-finished* task   a *partly finished* task

32c Hyphenate spelled-out compound numbers from *twenty-one* through *ninety-nine*.

32d Divide a word at the end of a line according to conventions.

Monosyllables

Do not divide.

thought   strength   cheese

Single letters

Do not put a one-letter syllable on a separate line.

NOT
a-bout   might-y

Prefixes and suffixes

May be divided.

> separ-able   pre-fix

Avoid carrying over a two-letter suffix.

> bound-ed   careful-ly

Compounds with hyphen

Avoid dividing and adding another hyphen.

> self-satisfied

NOT
> self-satis-fied

■ Exercise 3

Underline the correct form for the words indicated. Use a dictionary when needed.

1. (<u>Summertime</u>, Summer-time, Summer time) pleasure builds (<u>year-round</u>, year round) memories.
2. The (<u>thunderstorm</u>, thunder-storm, thunder storm) drove many people off the (golfcourse, golf-course, <u>golf course</u>) into the (justopened, <u>just-opened</u>, just opened) (<u>clubhouse</u>, club-house, club house).
3. The (<u>foxhound</u>, fox-hound, fox hound) was (welltrained, <u>well-trained</u>, well trained) not to chase rabbits.
4. The (twentyone, <u>twenty-one</u>, twenty one) dancers did not know how to (foxtrot, <u>fox-trot</u>, fox trot).
5. Several words in the paper were hyphenated at the end of lines: a-round, almight-y, <u>al-most</u>, self-in-flicted, marb-le.

33 Apostrophes ,

Use the apostrophe for the possessive case of many nouns, for contractions, for omissions, and for some plurals.

33a Use *'s* for the possessive of nouns not ending in *s*.

SINGULAR
child's, man's, deer's, lady's, mother-in-law's

PLURAL
children's, men's

33b Use *'s* for the possessive of singular nouns ending in *s*.

Charles's, Watts's, Dickens's, waitress's, actress's

NOTE: When a singular noun ending in *s* is followed by a word beginning with *s*, use only the apostrophe, not *'s*.

the actress' success, Dickens' stories

33c Use ' without *s* to form the possessive of plural nouns ending in *s*.

the Joneses' car, the Dickenses' home, waitresses' tips

33d   Use *'s* to form the possessive of indefinite pronouns.

anybody's, everyone's, somebody else's, neither's

NOTE: Use no apostrophe with personal pronouns like *his, hers, their, ours, its* (meaning "of it"). *It's* means "it is."

33e   Use *'s* with only the last noun for joint possession in a pair or a series.

the architect and the builder's plan [The two jointly have one plan.]
the architect's and the builder's plans [They have different plans.]

33f   Use *'* to show omissions or to form contractions.

o'clock, jack-o'-lantern
we'll, don't, can't, it's [meaning "it is"]

33g   Use *'s* to form the plural of acronyms and words being named.

three IRA's
six *the's*

NOTE: The plurals for numerals, letters, and years may be written with or without apostrophes as long as no confusion results (*U's* not *Us*, *i's* not *is*). Be consistent.

three *7's* or three 7s, four *C's* or four Cs, the 1960's or the 1960s, the Roaring '20's or the Roaring '20s

■ Exercise 4

Underline the words that contain correctly used apostrophes.

1. sheeps' wool, <u>deer's</u> horns, <u>cats'</u> eyes, a <u>cat's</u> eyes
2. the <u>youths'</u> organization, the <u>women's club,</u> the womens' club
3. the <u>childrens'</u> books, a childs' books, the two <u>boys'</u> books
4. the Beatles's songs, the Davises's vacation, <u>Dennis's</u> notebook
5. the sled that is her's, ours', <u>everyone's,</u> <u>somebody's</u> else
6. wasnt, <u>wasn't,</u> three *n*<u>'s</u>
7. three <u>*why*'s,</u> four <u>*how*s</u>
8. one <u>o'clock,</u> two oclock, three opossums
9. <u>Mary and Martin's</u> store (together they own one store)
   <u>Mary's and Martin's</u> stores (each owns a store)
10. our's, ours', its', <u>it's</u> (for *it is*)

34 Capital Letters *cap*

Use a capital letter to begin a sentence and to designate a proper noun (the name of a particular person, place, or thing).

34a Capitalize the first word of a sentence, the pronoun *I,* and the interjection *O.*

How, O ye gods, can I control this joy?

34b Capitalize first, last, and important words in titles, including the second part of hyphenated words.

> *Across the River and into the Trees*
> "The Man Against the Sky"
> "After Apple-Picking"
> "What You See Is the Real You"

NOTE: Articles *(a, an, the),* short prepositions, and conjunctions are not capitalized unless they begin or end a title.

34c Capitalize first words of direct quotations.

> The instructions warned, "Do not immerse in water."

34d Capitalize titles preceding a name.

> Professor Xavier

34e Capitalize the title of the head of a nation.

> The President of the United States is not expected to arrive today.

34f Capitalize titles used specifically as substitutes for particular names.

> Major Yo pleaded not guilty; the Major was found innocent.

NOTE: A title not followed by a name is usually not capitalized.

> The treasurer gave the financial report.

Titles that are common nouns that name an office are not capitalized.

A college president has more duties than privileges.
A lieutenant deserves a good living allowance.

34g Capitalize degrees and titles after a name.

Jeffrey E. Tyndale, Ph.D., J.D.
Sandra Day O'Connor, Associate Justice of the Supreme Court

NOTE: Do not capitalize names of occupations used as appositives or as descriptions.

Abraham Lincoln, a young lawyer from Springfield, took the case.

34h Capitalize words of family relationship used as names when not preceded by a possessive pronoun or the word *the*.

USED AS NAMES
After Father died, Mother carried on the business.

BUT
After my father died, my mother carried on the business.

34i Capitalize proper nouns but not general terms.

PROPER NOUNS	GENERAL TERMS
Plato, Platonic, Platonism	pasteurize
Venice, Venetian blind	a set of china
the West, a Westerner	west of the river
the Republican Party	a republican government
the Senior Class of Ivy College	a member of the senior class
Clifton Street	my street, the street
the Mississippi River	the Mississippi and Ohio rivers
the Romantic Movement	the twentieth century

34j   Capitalize months, days of the week, and holidays.

April, Friday, the Fourth of July, Labor Day

NOTE: Do not capitalize seasons and numbered days of the month unless they name holidays.

spring, the third of July

34k   Capitalize B.C. (used after numerals: 31 B.C.), A.D. (used before numerals: A.D. 33), words designating a deity, religious denominations, and sacred books.

in 273 B.C.
the Messiah, our Maker, the Trinity, Yahweh, Allah, Buddha, Jesus
"Praise God from Whom all blessings flow."
Catholic, Protestant, Presbyterian
the Bible, the Koran

NOTE: Pronouns referring to God are usually capitalized.

From Him all blessings flow.

34L   Capitalize names of specific courses.

I registered for Sociology 101 and Chemistry 445.

NOTE: Do not capitalize studies (other than languages) that do not name specific courses.

I am taking English, sociology, and chemistry.

35 Abbreviations *ab*

Avoid most abbreviations in formal writing.

35a Spell out names of days, months, units of measurement, and (except in addresses) states and countries.

Friday (*not* Fri.)	pounds (*not* lbs.)
February (*not* Feb.)	Sauk Centre, Minnesota (*not* Minn.)

Do not use note-taking or shortcut signs such as *w/* for *with* or *&* for *and* in formal writing.

35b Use only acceptable abbreviations.

BEFORE NAMES
  Mr., Mrs., Ms., Messrs., Mmes., Dr., St. *or* Ste. (for *Saint,* not *Street*), Mt., Rev. (but only with a first name: *the Rev. Ernest Jones,* not *Rev. Jones*)

AFTER NAMES
  M.D. (and other degrees), Jr., Sr., Esq.

FOR THE DISTRICT OF COLUMBIA
  Washington, D.C.

FOR MANY AGENCIES AND ORGANIZATIONS (WITHOUT PERIODS)
  TVA, NAACP, FBI, IRS, NBC

WITH DATES AND TIME
  B.C. and A.D. (with dates expressed in numerals, as *500 B.C.*), A.M. and P.M. or a.m. and p.m. (with hours expressed in numerals, as *4:00 A.M.* or *4:00 a.m.*)

36 Numbers   *num*

Spell out numbers or use numerals where appropriate.

36a Spell out numbers that can be written in one or two words. Use numerals for other numbers.

twenty-three, one thousand

123   1¹³⁄₁₆ $1,001.00

NOTE: Newspapers and government publications generally use numerals for numbers above ten.

EXCEPTION: Never use numerals at the beginning of a sentence. Spell out the number or recast the sentence.

36b Be consistent with numbers in a sequence.

NOT

> One polar bear weighed 426 pounds; another, 538 pounds; and the third, two hundred pounds.

BUT

> One polar bear weighed 426 pounds; another, 538 pounds; and the third, 200 pounds.

36c Use numerals for dates, street numbers, page references, percentages, and hours of the day used with A.M. or P.M.

USE NUMERALS	SPELL OUT
July 3, 1776 (*not* 3rd)	the third of July
1010 State Street	Fifth Avenue
See page 50.	The book has fifty pages.
He paid 15 percent interest.	
The concert begins at 6 P.M.	The concert begins at
(or 6:00 P.M.)	six o'clock.

■ Exercise 5

Supply capitals as needed below. Change capital letters to lowercase as necessary.

1. Dr. Outback, an **A**ustralian expert in **a**nimal **b**ehavior, lectures occasionally on **m**arsupial psychoses.

2. The **h**ostess, my **A**unt Zora, cried, "**H**elp yourself to the fried chicken," in a voice so shrill and strange that the dinner guests suddenly lost their appetite.

3. Captain Kaplan, **U**nited States **A**rmy, arrived on **W**ednesday to find that he was late for the tour of **B**uddhist temples.

4. When she registered for **c**hemistry, **A**maryllis was told that she would need to take Algebra 101.

5. Susan Curall, **M.D.**, attended the meeting of the American Medical **A**ssociation and returned home before **T**hanksgiving **D**ay.

6. In the ^tTwentieth ^cCentury, many of the qualities associated
with people of the American ^SSouth have disappeared.

7. Augustus Caesar, who was born in 63 <del>b.c.</del> **B.C.** and died in
<del>a.d.</del> **A.D.** 14, is a character in Shakespeare's ^tTragedy *Antony and*
Cleopatra.

8. Though the printer lived for a while on Magoni ^AAvenue, he
moved to ^Ddetroit last August.

9. The Salk ^vVaccine has all but eliminated ^pPolio, according to
an article in a ^mMedical ^jJournal.

10. She wanted to become a ^lLawyer, she explained, because she
saw a direct connection between the ^lLaw and ^mMorals.

■ Exercise 6

Place an X *by the following that are not acceptable in formal*
writing, a ✔ *by those that are acceptable.*

✔   1. May thirteenth
X   2. Twelve thirteen Jefferson Street
X   3. Main St.
X   4. Mister and Mrs. Smidt
X   5. seven P.M.
✔   6. six thousand votes
X   7. Minneapolis, Minn., on 3 June 1940

X 8. Eng. 199 in the Dept. of English
✓ 9. page 10 in Chapter 11
X 10. Chas. Lorenzo
X 11. Alfred Ginsberg, Junior
✓ 12. 300 B.C.
X 13. Friday and Saturday, May 16 & 17
X 14. 6 pounds, 7 oz.
X 15. five hundred and ten bushels
X 16. December 24th, 1985
✓ 17. $4.98 a pound
X 18. handle w/ care
✓ 19. five million dollars
✓ 20. Friday, June 13

Diction
and Style

37 Diction   *d*

Select words that conform to established usage. Use Standard English appropriate for the occasion.

Diction means choice and use of words. **Standard English** is the generally accepted language of educated people in English-speaking countries. Though it varies in usage and in pronunciation from one country or region to another (indicated in dictionaries by labels such as *U.S.* or *Brit.*), it is the standard taught in schools and colleges.

Nonstandard English consists of usages, spellings, and pronunciations not usually found in the speech or writing of educated people. Read the usage labels in a dictionary to determine whether a word is nonstandard.

Informal or **colloquial English** reflects the spoken language and is used in writing that seeks the effect of casual or everyday speech. It is usually not appropriate in college papers. *Colloquial* does not mean *dialect* or *local language.*

NOTE: **Contractions**—*don't, isn't,* and so on—are informal.

37a   Using a dictionary   *dic*

Dictionaries, which are good sources of information about language, record current and past usage. In minor matters they do not always agree.

Most brief paperback dictionaries provide too little information to be appropriate for college students. Particularly useful at the college level are the following desk dictionaries:

The American Heritage Dictionary. Boston: Houghton Mifflin.

The Random House College Dictionary. New York: Random House.

Webster's New Collegiate Dictionary. Springfield, Mass.: Merriam.

Webster's New World Dictionary of the American Language. Cleveland: William Collins.

Webster's II: New Riverside University Dictionary. Boston: Houghton Mifflin.

Know how to look up a word, a word group, and many other kinds of information in your dictionary. You do not need to know all the different methods used by different dictionaries. Instead, select a good one and study carefully its explanations and systems. Read the entry on p. 195 for the word *double* as given in *The American Heritage Dictionary,* and study the explanatory material in red. Use this example as a guide. (A similar chart is included in the prefatory material of *Webster's New Collegiate Dictionary.*)

Dictionary entry

Dictionaries provide the kinds of information indicated here in several ways. Other information (synonyms, variants, capitalizations, and so on) appears in many entries. Find the explanations you need in any dictionary.

① Boldfaced main entry ② Dot indicating division between syllables ③ Phonetic spelling for pronunciation (dictionaries place a pronunciation key at the bottom of each page) ④ Accented syllable ⑤ Abbreviation for part of speech—adjective ⑥ Illustrative example ⑦ Words or abbreviations in red blocks indicate terms that label specialized or technical meanings. ⑧ Part of speech—noun. Note that the numbers start over when the part of speech changes. ⑨ Part of speech—verb ⑩ Forms of the verb—past tense and past participle, present participle, third-person singular ⑪ Kind of verb—transitive (see section **5** in this book) ⑫ Kind of verb—intransitive (see section **5** in this book) ⑬ Phrasal verb (used in *AHD* to mean verb plus adverb *or* preposition) ⑭ Part of speech—adverb ⑮ Idioms (see p. 199 in this book) ⑯ Usage label (see p. 192 in this book) ⑰ Etymologies are given in square brackets. Some dictionaries give the etymology early in the entry. ⑱ A noun derived from the main entry

dou·ble (dŭb′əl) *adj.* **1.** Twice as much in size, strength, number, or amount: *a double dose.* **2.** Composed of two like parts: *double doors.* **3.** Composed of two unlike parts; dual: *a double meaning.* **4.** Accommodating or designed for two: *a double sleeping bag.* **5. a.** Acting two parts: *a double role.* **b.** Characterized by duplicity; deceitful: *speak with a double tongue.* **6.** *Bot.* Having many more than the usual number of petals, usually in a crowded or overlapping arrangement: *a double chrysanthemum.* —*n.* **1.** Something increased twofold. **2. a.** A duplicate of another; counterpart. **b.** An apparition; wraith. **3.** An actor's understudy. **4. a.** A sharp turn in running; reversal. **b.** An evasive reversal or shift in argument. **5. doubles.** A game, such as tennis or handball, having two players on each side. **6.** *Baseball.* A two-base hit. **7. a.** A bid in bridge indicating strength to one's partner; request for a bid. **b.** A bid doubling one's opponent's bid in bridge thus increasing the penalty for failure to fulfill the contract. **c.** A hand justifying such a bid. —*v.* **-bled, -bling, -bles.** —*tr.* **1.** To make twice as great. **2.** To be twice as much as. **3.** To fold in two. **4.** To duplicate; repeat. **5.** *Baseball.* **a.** To cause the scoring of (a run) by hitting a double. **b.** To advance or score (a runner) by hitting a double. **6.** *Baseball.* To put out (a runner) as the second part of a double play. **7.** To challenge (an opponent's bid) with a double in bridge. **8.** *Mus.* To duplicate (another part or voice) an octave higher or lower or in unison. **9.** *Naut.* To sail around: *double a cape.* —*intr.* **1.** To be increased twofold. **2.** To turn sharply backward; reverse: *double back on one's trail.* **3.** To serve in an additional capacity. **4.** To replace an actor in the execution of a given action or in the actor's absence. **5.** *Baseball.* To hit a double. **6.** To announce a double in bridge. —*phrasal verb.* **double up. 1.** To bend suddenly, as in pain or laughter. **2.** To share accommodations meant for one person. —*adv.* **1. a.** To twice the extent; doubly. **b.** To twice the amount: *double your money back.* **2.** Two together: *sleeping double.* **3.** In two: *bent double.* —**idioms. on** (or **at**) **the double.** *Informal.* **1.** In double time. **2.** Immediately. **see double.** To see two images of a single object, usually as a result of visual aberration. [ME < OFr. < Lat. *duplus.*] —**dou′ble·ness** *n.*

■ Exercise 1

Read the preliminary pages in your dictionary, and study the labels (such as Mus., Math., Nonstandard, Informal, Slang, Vulgar, Regional, Chiefly Brit., *and so forth) that it applies to particular words.* Webster's New Collegiate Dictionary *uses fewer and less restrictive labels than other college dictionaries. Without using your dictionary, put by ten of the following words the labels, if any, that you believe they should have. Then look up each word, and determine which label it has. If a word is not labeled, it is considered formal.*

1. flat out (at top speed)
2. deck (data-processing cards)
3. gussied up
4. on deck (present)
5. movie
6. ain't
7. cram (to gorge with food)
8. tube (subway)
9. tube (television)
10. phony
11. bash (a party)
12. bash (a heavy blow)
13. enthused
14. pigeon (a dupe)
15. balloon (to rise quickly)
16. freak (an enthusiast)
17. squeal (to betray)
18. deck (a tape deck)
19. wimp
20. on the level

37b Generally, avoid slang. *sl*

Slang expressions in student papers are usually out of place. They are particularly inappropriate in a context that is otherwise dignified.

> In the opinion of many students, the dean's commencement address *stunk.*

> When Macbeth recoiled at the thought of murder, Lady Macbeth urged him not to *chicken out.*

In addition, some slang words are so recent or localized that many people will have no idea of the meaning.

Although slang can be out of place, puzzling, or even offensive, it is at times effective. Some words that once were slang are now considered Standard English. "Jazz" in its original meaning and "dropout" were once slang; and because no other word was found to convey quite the same social meanings as "date," it is no longer slang. As a rule, however, avoid words still considered slang, not only because they are too modern for conservative ears and eyes or because they are offensive to stuffy people, but also because they are generally not as precise or as widely known as Standard English.

37c Avoid dialect.   *dial*

Regional, occupational, or ethnic words and usages should be avoided in formal writing except when they are consciously used to give a special flavor to language.

There is no reason to erase all dialectal characteristics from talk and writing. They are a cultural heritage and a source of richness and variety.

37d Avoid archaic words.   *archaic*

Archaisms (*oft* for "frequent" and *yond* for "yonder") are out-of-date words and therefore inappropriate in modern speech and writing.

■ Exercise 2

In class, discuss usage in the following sentences.

1. If someone is putting you down, remain calm and simply tell him or her to chill out.
2. Why get bent out of shape over nothing? Play it cool.
3. That green stuff we had for dinner was really gross.
4. When the young stockbroker discovered the potential earning power of the mining company, he freaked out.

5. He couldn't do nothing without the supervisor eyeballing him and giving him a hard time.
6. The speeders heard on their ears that smokey was only three miles away on the blacktop.
7. The teller of the bank was instructed to tote the money back to the vault and to follow the boss to his office.
8. What cleared him was that the feds found out who really done the robbery.
9. Ere the rain ceased, the tourists were besprinkled, but their enthusiasm was not dampened.
10. The dispatcher reckoned that the semi would arrive betimes, but he could not raise the driver on the radio.

37e Do not use a word as a particular part of speech when it does not properly serve that function.

Many nouns, for example, cannot also be used as verbs and vice versa. When in doubt, check the part of speech of a word in a dictionary.

NOUN FOR ADJECTIVE *occupation* hazard	VERB DERIVED FROM NOUN FORM *suspicioned*
VERB FOR NOUN good *eats*	VERB FOR ADJECTIVE *militate* leader
ADJECTIVE FOR ADVERB sing *good*	

For words that are used inexactly in meaning (rather than function), see **37g**.

37f   Use correct idioms.   *id*

An idiom is an accepted expression with its own distinct form and meaning. Every language is rich in idiomatic expressions, some of which do not seem to make much sense literally but nevertheless have taken on specific meanings with time.

> An onyx ring in the jewelry store *caught her eye*.
>
> Though the gale was fierce, there was nothing else to do but *ride it out*.
>
> The travel agent indicated that the ship would sail *Friday week*.

Many expressions demand specific prepositions; unidiomatic writing results when prepositions are misused.

UNIDIOMATIC	IDIOMATIC
according with	according to
capable to	capable of
conform in	conform to (*or* with)
die from	die of
ever now and then	every now and then
excepting for	except for
identical to	identical with
in accordance to	in accordance with
incapable to do	incapable of doing
in search for	in search of
intend on doing	intend to do
in the year of 1976	in the year 1976
lavish with gifts	lavish gifts on
off of	off
on a whole	on the whole
outlook of life	outlook on life
plan on	plan to
prior than	prior to
similar with	similar to
superior than	superior to
try and see	try to see
type of a	type of

Dictionaries list many idiomatic combinations and give their exact meanings. See ⑮ on pp. 194–195.

■ Exercise 3

*Select a verb that is the common denominator in several idioms,
such as* carry (carry out, carry over, carry through, *and so forth*).
*Look it up in a dictionary. Study the idioms and be prepared to
discuss them in class. Does the change of a preposition result in
a different meaning? A slight change in meaning? No change?*

■ Exercise 4

*Underline the incorrect prepositions and supply the correct ones
in the following sentences. Write* C *by correct sentences.*

1. After studying with a tutor, the student who was behind in

   with
   algebra was able to catch up to the rest of the class.

   of
2. The jury acquitted him for the charge of loitering.

   to
3. The aging baritone reacted favorably with the suggestion that

   he retire and surprised all members of the opera company.

   of
4. The hiker said that he was incapable to going on.

5. The professor stated that his good students would conform
   to
   with any requirement.

   to
6. The employees complained with the manager because of

   their long hours.

   in
7. Many vitamins are helpful to preventing diseases.

C   8. Fourteen new people were named to posts in the university administration.

On
9. Upon the whole, matters could be much worse.

             in
10. Let us rejoice for the knowledge that we are free.

37g Use words in their precise meanings, and select words with appropriate connotations.   *con*

Denotations of words are their precise meanings, the exact definitions given in dictionaries. In addition, many words carry special associations, suggestions, or emotional overtones—**connotations.**

Denotation

Misuse of a word so that it does not convey the intended meaning produces confusion. Check a dictionary whenever you have the slightest question about a word's denotation. The italicized words in the following sentences are misused.

> The new rocket was *literally* as fast as lightning. [*Figuratively* is intended.]
> The captain of the team was *overtaken* by the heat. [The proper word is *overcome.*]

Words that are similar in sound but different in meaning can cause the writer embarrassment and the reader confusion.

> A wrongful act usually results in a guilty *conscious.* [*conscience*]
> The heroine of the novel was not embarrassed by her *congenial* infirmity. [*congenital*]

As the sun beams down, the swamp looks gray; it has only the unreal color of dead *vegetarian*. [*vegetation*]

A hurricane *reeks* havoc. [*wreaks*]

Displaced persons sometimes have no *lengths* with the past. [*links*]

Words sometimes confused because they sound alike are *climatic* for *climactic,* *statue* for *stature* (or vice versa), *incidences* for *incidents,* and *course* for *coarse.* Nonwords should never be used: *interpretate* for *interpret,* and *tutorer* for *tutor.*

Sometimes a fancy word seems effective when it is actually inexact in meaning, such as "scenario of the poem" for "story told in the poem." *Scenario* more exactly means "outline."

Inexactness and ambiguity result when a word in a phrase can be perceived in either of two ways; for instance, "perception of the judge" can refer to the understanding *by* the judge of the case or to the understanding *of* the judge by a lawyer.

Connotation

A good writer selects words with connotations that evoke planned emotional reactions. To suggest sophistication, the writer may mention a *lap dog;* to evoke the amusing or the rural, *hound dog.* To make a social or moral judgment, the writer may call someone a *cur.* Consider the associations aroused by *canine, pooch, mutt, mongrel, puppy,* and *watchdog.* Even some breeds arouse different responses: *bloodhound, shepherd, St. Bernard, poodle.*

Words that are denotative synonyms may have very different connotative overtones. Consider the following:

drummer—salesperson—field representative

slender—thin—skinny

Be sure that your words give the suggestions you wish to convey. A single word with the wrong connotation can easily spoil a passage. President Lincoln once said in a speech:

Human nature will not change. In any future great national trial, compared with the men of this, we shall have as weak and as strong, as silly and as wise, as bad and as good.

Notice how substituting only two words ruins the consistent high tone of the passage, even though the meanings of the substitutes are close to those of the original.

Folks will not change. In any future great national trial, compared with the men of this, we shall have as weak and as strong, as silly and as wise, as crooked and as good.

■ Exercise 5

Underline vague words or words used incorrectly in the following sentences, and make necessary substitutions.

> **lush**
1. A luscious carpet covered the floor of the restaurant.

>   **delicious**   **impeccable**
2. The meal was simply fabulous and the service unbelievable—
>   **excellent**
   all in all, fantastic.

>   **elegy**
3. Gray's allergy, a truly moving poem, was written in a time of
>   **mourning**
   great change and morning for the past.

4. On the outskirts of the small town a sign was erected that
>   **announced**
   renounced that this was the home of Marlin Fish.

>   **baffling**
5. Out there in space are many babbling questions to be

   answered.

6. Americans obtain citizenship either by being born in this
   naturalized
   country or by becoming nationalized.
   Recounting
7. Accounting all the events that led to the fall of Rome would

   be impossible in a single essay.
   keynote
8. Adrian Womack will be the keystone speaker tonight.
   moot
9. "That is a mute point," shouted the attorney. "Besides, it is
   irrelevant
   irrevalent and immaterial."

10. At the retirement dinner the bank president mentioned that
    presence   **sorely**
    the teller's daily presents would be soarly missed.

■ Exercise 6

*The ten words or phrases in the left column below name a subject.
The right column gives a word referring to that subject. For each
word in the right column, list a close synonym that is much more
favorable in connotation and another that is less favorable.*

1. weight	stout
2. intelligence	sense
3. writing style	intelligible
4. food	edible
5. degree of value	economical
6. personality	fairly agreeable
7. physical skill	coordinated
8. exactness of measurement	close enough
9. efficiency	competent
10. beauty	attractive

37h Avoid specialized vocabulary in writing for the general reader. *tech*

All specialists, whether engineers, chefs, or philosophers, have their own vocabularies. Some technical words find their way into general use; most do not. The plant red clover is well known but not by its botanical name, *Trifolium pratense.*

Specialists should use the language of nonspecialists when they hope to communicate with general readers. The following passage, for instance, would not be comprehensible to a wide audience.

> The neonate's environment consists in primitively contrasted perceptual fields weak and strong: loud noises, bright lights, smooth surfaces, compared with silence, darkness and roughness. The behavior of the neonate has to be accounted for chiefly by inherited motor connections between receptors and effectors. There is at this stage, in addition to the autonomic nervous system, only the sensorimotor system to call on. And so the ability of the infant to discriminate is exceedingly low. But by receiving and sorting random data through the identification of recurrent regularity, he does begin to improve reception. Hence he can surrender the more easily to single motivations, ego-involvement in satisfactions.
>
> JAMES K. FEIBLEMAN
> *The Stages of Human Life: A Biography of Entire Man*

Contrast the above passage with the following, which is on the same general subject but which is written so that the general reader—not just a specialized few—can understand it.

> Research clearly indicates that an infant's senses are functional at birth. He experiences the whack from the doctor. He is sensitive to pressure, to changes in temperature, and to pain, and he responds specifically to these stimuli. . . . How about sight? Research on infants 4–8 weeks of age shows that they can see about as well as adults.

. . . The difference is that the infant cannot make sense out of what he sees. Nevertheless, what he sees does register, and he begins to take in visual information at birth. . . . In summary, the neonate (an infant less than a month old) is sensitive not only to internal but also to external stimuli. Although he cannot respond adequately, he does take in and process information.

IRA J. GORDON
Human Development: From Birth Through Adolescence

The only technical term in the passage is *neonate;* unlike the writer of the first passage, who also uses the word, the second author defines it for the general reader.

Special vocabularies may obscure meaning. Moreover, they tempt the writer to use inflated words instead of plain ones—a style sometimes known as *gobbledygook* or *governmentese* because it flourishes in bureaucratic writing. Harry S. Truman made a famous statement about the presidency: "The buck stops here." This straightforward assertion might be written by some bureaucrats as follows: "It is incumbent upon the President of the United States of America to uphold the responsibility placed upon him by his constituents to exercise the final decision-making power."

37i   Add new words to your vocabulary.   *vocab*

Good writers know many words, and they can select the precise ones they need to express their meanings. A good vocabulary displays your mentality, your education, and your talents as a writer.

In reading, pay careful attention to words you have not seen before. Look them up in a dictionary. Remember them. Recognize them the next time you see them. Learn to use them.

■ Exercise 7

Underline the letter identifying the best definition.

1. *contingency:* (a) series   (b) important point   (c) possible condition   (d) rapidly
2. *aegis:* (a) sponsorship   (b) eagerness   (c) foreign   (d) overly proper
3. *summit:* (a) highest point   (b) total amount   (c) exhibition (d) heir
4. *magnanimity:* (a) state of wealth   (b) excellent health (c) strong attraction   (d) generosity
5. *credibility:* (a) debt   (b) kindness   (c) knowledge   (d) believability
6. *gullible:* (a) capable of flight   (b) flexible   (c) easily tricked (d) intelligent
7. *patent:* (a) obvious   (b) long-suffering   (c) omen   (d) victim
8. *audible:* (a) capable of being heard   (b) overly idealistic (c) easily read   (d) tasty
9. *adamant:* (a) without measure   (b) unyielding   (c) fragrant (d) judicial decision
10. *travesty:* (a) long journey   (b) congested traffic   (c) scaffolding (d) mockery
11. *innate:* (a) void of sense   (b) inborn   (c) applied   (d) digestible
12. *acrimonious:* (a) bitter   (b) ritualistic   (c) hypocritical (d) massive
13. *insurgents:* (a) deep cuts   (b) rebels   (c) music makers (d) physicians
14. *duress:* (a) fancy dress   (b) with quickness   (c) stress (d) penalty
15. *flagrant:* (a) odorous   (b) conspicuously bad   (c) beaten (d) delicious

16. *substantive:* (a) substantial (b) in place of (c) religious (d) hardheaded

17. *imprudent:* (a) unwise (b) lacking modesty (c) ugly (d) incapable of proof

18. *solicitous:* (a) seeking sales (b) without energy (c) heroic (d) concerned

19. *valid:* (a) butler (b) favorable (c) desirable (d) founded on truth

20. *charlatan:* (a) robe (b) quack (c) high official (d) strong wind

■ Exercise 8

Underline the letter identifying the best definition.

1. *impetus:* (a) egomaniac (b) incentive (c) ghost (d) perfectionist

2. *dire:* (a) terrible (b) bare (c) final (d) dishonest

3. *subpoena:* (a) below ground (b) in disguise (c) legal summons (d) unspoken

4. *commentary:* (a) explanations (b) military store (c) grouch (d) headland

5. *prodigy:* (a) extraordinary person (b) wasteful person (c) lover of children (d) musician

6. *vacillate:* (a) oil (b) repair (c) waver (d) empty

7. *peer:* (a) an equal (b) an ideal (c) an emotion (d) intense

8. *poignant:* (a) housecoat (b) fruitful (c) hostile (d) piercingly effective

9. *spurious:* (a) brimming over (b) pricking (c) not genuine (d) affectionate

10. *perennial:* (a) continuing (b) circus show (c) circular (d) filial

11. *contend:* (a) incantate (b) appease (c) compete (d) look after

12. *scapegoat:* (a) one being sought (b) one taking blame for others (c) one who looks stupid (d) one guilty of crime

13. *pomposity:* (a) heaviness (b) splendor (c) scarcity (d) self-importance

14. *banal:* (a) commonplace (b) fatal (c) aggressive (d) tropical

15. *parameter:* (a) restatement (b) expansion (c) limit (d) device for measuring sound

16. *acute:* (a) appealing (b) severe (c) average (d) unorthodox

17. *vibrant:* (a) blunted (b) double (c) irritating (d) energetic

18. *presume:* (a) guess (b) judge (c) know (d) take for granted

19. *gesture:* (a) demonstration (b) internal disorder (c) joke (d) court clown

20. *veracity:* (a) boldness (b) speed (c) great anger (d) truthfulness

■ Exercise 9

Place the number on the left by the appropriate letter on the right.

1. heady	**12**	a.	exaggeration	
2. reciprocate	**13**	b.	customary	
3. dispiriting	**2**	c.	repay	
4. literally	**17**	d.	a ranking	
5. consecrate	**15**	e.	imitation	
6. bolster	**16**	f.	basic	
7. ideology	**4**	g.	in actuality	
8. anticipate	**10**	h.	unreal	
9. incorrigible	**11**	i.	opposition	
10. fantastic	**1**	j.	impetuous	
11. antagonism	**14**	k.	a showing forth	
12. hyperbole	**18**	l.	one who comes before	
13. conventional	**19**	m.	twist	
14. manifestation	**9**	n.	not reformable	
15. mimicry	**3**	o.	disheartening	
16. seminal	**8**	p.	foresee	
17. hierarchy	**5**	q.	make sacred	
18. predecessor	**7**	r.	set of beliefs	
19. distort	**20**	s.	maze	
20. labyrinth	**6**	t.	support	

■ **Exercise 10**

Underline the word or phrase that most exactly defines the italicized word in each sentence.

1. An *obdurate* attitude seldom leads to prosperity. (egotistical, hardhearted, wasteful, false)
2. He felt it his *prerogative* to speak out. (turn, responsibility, right, nature)
3. The speech was marked by *vapid* expressions. (inspiring, dull, vigorous, sad)
4. The *renowned* singer performed in the small town. (famous, unknown, opera, talented)
5. *Gluttony,* explained the slim traveler, was not one of his faults. (hitchhiking, stagnation, overeating, speeding)
6. The banker's *salient* traits were frugality and kindness. (prominent, best, hidden, lacking)
7. Imagining himself a great statesman, he was the prime minister's *lackey*. (assistant, servile follower, hair groomer, moral superior)
8. For the greater part of a lengthy career, the poet was *impecunious*. (unpublished, unpopular, poor, peculiar)
9. Not all those who took part in the robbery were *depraved*. (needy, evil, prepared, punished)
10. The gem was pronounced an *authentic* emerald. (genuine, artificial, rare, expensive)

■ **Exercise 11**

Follow the instructions for Exercise 10.

1. The judge found the defendant's answer *incredible*. (highly impressive, wonderful, not believable, awful)

2. A *transcript* was made of the tapes. (recording, written copy, mockery, extension)
3. To discuss the matter further would be to *obfuscate* it. (avoid, obscure, criticize, clarify)
4. I *deplore* the method used to recover the gems. (praise, understand, regret, follow)
5. The poem was composed by an *anonymous* author. (dead, foreign, excellent, unknown)
6. Finding the essay *provocative,* she discussed it with her tutor. (stimulating, disgusting, prolonged, offensive)
7. The child *reluctantly* joined her brother in the swimming pool. (rapidly, hesitantly, joyfully, playfully)
8. *Platitudes* can quickly destroy the effectiveness of a lecture. (catcalls, stutterings, stale truisms, bad jokes)
9. A *sorcerer* was said to be the king's only companion. (healer, valiant warrior, jester, wizard)
10. Make the report as *succinct* as possible. (colorful, concise, long, accurate)

38 Style   *st*

Express yourself with economy, clarity, and freshness.

Style refers to the way writers express their thoughts in language, the way they put words together. The modes in which similar ideas are expressed can have different effects, and many of the distinctions are a matter of style. Acquire the habit of noticing the personality of what you read and write, and work to develop your own style.

38a Avoid verbosity; omit needless words.   *w*

Conciseness increases the force of writing. A verbose style is flabby and ineffectual. Do not pad your paper merely to obtain a desired length.

USE ONE WORD FOR MANY

The love letter was written by somebody who did not sign a name. [13 words]

The love letter was anonymous [*or* not signed]. [5 or 6 words]

USE THE ACTIVE VOICE FOR CONCISENESS (See **5**)

The truck was overloaded by the workmen. [7 words]

The workmen overloaded the truck. [5 words]

REVISE SENTENCES FOR CONCISENESS

Another element that adds to the effectiveness of a speech is its emotional appeal. [14 words]

Emotional appeal also makes a speech more effective. [8 words]

AVOID CONSTRUCTIONS WITH *IT IS* . . . AND *THERE ARE* . . .

It is truth *that* will prevail. [6 words]

Truth will prevail. [3 words]

There are some conditions *that* are satisfactory. [7 words]

Some conditions are satisfactory. [4 words]

DO NOT USE TWO WORDS WITH THE SAME MEANING (tautology)

basic and fundamental principles [4 words]

basic principles [2 words]

Study your sentences carefully, and make them concise by using all the preceding methods. Do not, however, sacrifice concreteness and vividness for conciseness and brevity.

CONCRETE AND VIVID
> At each end of the sunken garden, worn granite steps, flanked by large magnolia trees, lead to formal paths.

EXCESSIVELY CONCISE
> The garden has steps at both ends.

■ Exercise 12

Express the following sentences succinctly. Do not omit important ideas.

1. The kudzu plant is a plant that was introduced into America from Japan in order to prevent erosion and the washing away of the land and that has become a nuisance and a pest in some areas because it chokes out trees and other vegetation.
   Kudzu, which was introduced into America from Japan to pre-vent erosion, has become a nuisance in some areas because it chokes out other plants.

2. The cry of a peacock is audible to the ear for miles.
   The cry of a peacock is audible for miles.

3. The custom that once was so popular of speaking to fellow students while passing by them on the campus has almost disappeared from college manners and habits.
   The once-popular custom of speaking to fellow students on the campus has almost disappeared.

4. There are several reasons why officers of the law ought to be trained in the law of the land, and two of these are as follows. The first of these reasons is that police officers can enforce the law better if they are familiar with it. And second, they will be less likely to violate the rights of private citizens if they know exactly and accurately what these rights are.
   Officers should know the law so that they can enforce it better and avoid violating the rights of citizens.

5. Although the Kentucky rifle played an important and significant part in getting food for the frontiersmen who settled the American West, its function as a means of protection was in no degree any less significant in their lives.

   Although the Kentucky rifle played an important part in getting food for the frontiersmen, its protective function was just as significant.

6. Some television programs, especially public television programs, assume a high level of public intelligence and present their shows to the public in an intelligent way.

   Some public television programs assume an intelligent audience and present their shows accordingly.

7. It is possible that some of the large American chestnut trees survived the terrible blight of the trees in the year of 1925.

   Possibly some large American chestnut trees survived the terrible blight of 1925.

8. The pilot who will fly the plane is a cautious man, and he is careful in every way.

   The pilot is a cautious flyer.

9. It is a pleasure for some to indulge in eating large quantities of food at meals, but medical doctors of medicine tell us that such pleasures can only bring with them unpleasant results in the long run of things.

   The joys of overeating, doctors say, can have unpleasant results.

10. The essay consists of facts that describe vividly many aspects of the work of a typical stockbroker. In this description the author uses a vocabulary that is easy to understand. This vocabulary is on neither too high a level nor too low a level, but on one that can be understood by any high school graduate.

    In simple language the essay describes vividly many aspects of the work of a typical stockbroker.

38b Avoid repeating ideas, words, or sounds.   *rep*

Unintentional repetition is a mark of bad style. Avoid by omitting words and by using synonyms and pronouns.

REPETITIOUS

The history of human flight is full of histories of failures on the part of those who have tried flight and were failures.

IMPROVED

The history of human flight recounts many failures.

Do not revise by substituting synonyms for repeated words too often.

WORDY SYNONYMS

The history of human flight is full of stories of failures on the part of those who have tried to glide through the air and met with no success.

Do not needlessly repeat sounds.

REPEATED SOUNDS

The biologist again *checked* the *charts* to determine the *effect* of the poison on the *insect.*

IMPROVED

The biologist again studied the charts to determine the effect of the poison on the moth.

NOTE: Effective repetition of a word or a phrase may unify, clarify, or create emphasis, especially in aphorisms or poetry.

Searching without knowledge is like *searching* in the dark.
"Beauty is truth, truth beauty." [John Keats]

■ Exercise 13

Rewrite the following passage. Avoid wordiness and undesirable repetition.

P **and ages**
~~One of the pleasing things that all~~ peoples of all lands have

always enjoyed ~~since the earliest dawning of civilizations is the~~

, which,
~~pleasure of listening to the~~ melodious strands of music, ~~Music,~~

some great poet has said,   **.**
~~wrote some great poet,~~ can soothe the savage beast ~~and make~~

Some younger people, however, may be listening
~~him or her calm. A question may arise in the minds of many,~~
too frequently to their favorite music.
~~however, as to whether some of our members of the younger~~

~~generation may not be exposing themselves too frequently to~~

~~music they like and listen to almost constantly.~~

Nearly every day, for example, one meets a young person who lis-
~~To me, the answer to the above question is just possibly in~~
tens to music
~~the affirmative, and the evidence that I would give would be in~~

~~the form of a figure that we see nearly every day. This is the~~

~~figure, usually a young person but not always, of a person with a~~

~~radio or stereo headset who is listening to music~~ while doing an
                     should   **his or her full**   **.**   **At**
activity that ~~used to~~ occupy ~~all of one's~~ attention ~~and time. In the~~
times a person in the library looks through books while wearing a
~~library of a college such a figure is now at times to be seen looking~~

headset.

~~through books with a headset on his or her head.~~ Some ~~of these~~
      of this sort   **and enjoying**
music lovers seem incapable of even taking a walk ~~to enjoy~~ the

beauty of nature without their headsets.

   If we enjoy good music,
   ~~If music is good and if we enjoy it, the reader may now ask,~~
                 it as   **as modern**
what is wrong with listening to ~~music more~~ frequently ~~than we~~
electronics enables us to do? It is not old-fashioned to believe that
~~used to be able to listen to music thanks to the present modern~~
those who spend much of their time with headsets on are missing
~~improvements and developments in the electronic medium? It~~
other important and pleasurable sensory experiences—the sounds of
~~may seem old-fashioned and out-of-date to object to headsets just~~
birds and even the quiet of morning. Further, it is hard to think while
~~because we did not once have them and they are relatively new~~
the ear is piped full of music. Time that should be used for thinking
~~on the scene. What is important, here, however, is the point that~~
is also lost—time when we develop and mature intellectually. All
~~those who spend so much of their time with headsets on are~~
kinds of music, even rock, have an important place, but music should
~~missing something. They may be getting a lot of music, but they~~
not be allowed to exclude too many other valuable experiences.
~~are missing something. What they are missing can be classified~~

~~under the general heading of other important and pleasurable~~

~~sensory experiences that they could be experiencing, such as~~

~~hearing the sounds that birds make and just enjoying the quiet~~

~~that morning sometimes brings. In addition to this, it is hard to~~

~~think while music is being piped into the ear of the listener. We~~

~~lose time, therefore, that should by all rights be reserved for~~
~~thinking and contemplation, which is to say, that time when we~~
~~all develop and mature intellectually. Music, all kinds of music,~~
~~even rock music, has an important place in the lives of humankind,~~
~~but let us not so fall in love with its seductive appeal that beckons~~
~~us that we let it intrude upon the territory of other valuable and~~
~~essential experiences.~~

38c   Avoid flowery language.   *fl*

Flowery language is wordy, overwrought, and artificial. Often falsely elegant, it calls attention to itself. In the hope that such language will sound deep and wise, some inexperienced writers substitute it for naturalness and simplicity.

PLAIN LANGUAGE	FLOWERY LANGUAGE
the year 1981	the year of 1981
now	at this point in time
lawn	verdant sward
shovel	simple instrument for delving into Mother Earth
a teacher	a dedicated toiler in the arduous labors of pedagogy
reading a textbook	following the lamp of knowledge in a textual tome
eating	partaking of the dietary sustenance of life
going overseas	traversing the ever-palpitating deep

38d   Be clear.   *cl*

A hard-to-read style annoys and alienates. Do not take for granted that others will be patient or understanding enough to pore over your writing to determine your meaning. If you have any doubt that a sentence or a passage conveys the precise meaning you intend, revise and clarify. Nothing is more important in writing than making sense. Below are a few suggestions about communicating more clearly.

Write specifically.

Abstract writing may communicate little information or even cause misunderstanding. Specific words say what you think. Their meaning is not ambiguous.

NOT SPECIFIC
> There was always a certain something in her personality that created a kind of positive effect.

SPECIFIC
> Her optimism always conveyed a sense of hope and joy.

Give concrete examples.

Concrete examples often pierce to the heart of meaning, whereas vague and abstract writing ineffectually ranges around the periphery.

CONCRETE
> Her bright, quick smile and her musical way of saying "Good morning!" conveyed a sense of hope and joy.

■ Exercise 14

The following ten words or phrases are general. For any five of them substitute four specific and concrete words. Do not let your substitutions be synonymous with each other.

EXAMPLE
    tools   claw hammer, screwdriver, monkey wrench, saw

1. associations
2. rural (or urban) buildings
3. people who are failures
4. works of art
5. terrains
6. nuts (or vegetables or meats or breads)
7. missives
8. educational institutions above the second year of college
9. foot races
10. ways to walk

■ Exercise 15

Write your personal definition of one of the following abstract terms in a paragraph of about two hundred words. Give concrete instances from your experience.

1. education
2. family
3. maturity
4. pride
5. era

38e Avoid triteness and clichés. Strive for fresh and original expressions.   *trite*

Clichés are phrases and figures of speech that were once fresh and original but have been used so much that they have lost their effectiveness. Avoid extravagance, but be original enough so that your words have the freshness of a newly typed page rather than the faint tracings of a carbon copy.

Study the following twenty phrases as examples of triteness. Avoid pat expressions like these:

words cannot express	method in their madness
each and every	straight from the shoulder
Mother Nature	first and foremost
sober as a judge	hard as a rock
other side of the coin	in the final analysis
slowly but surely	felt like an eternity
in this day and age	the bottom line
few and far between	all walks of life
last but not least	easier said than done
interesting to note	better late than never

■ Exercise 16

Underline the clichés in the following sentences.

1. Once in a blue moon, Professor Alhambra tells a joke that is funny, but then he spoils it by laughing like a hyena.
2. International Bugging Machines is on the cutting edge of technology; therefore, its stock is expected to soar through the roof.
3. It goes without saying that the value of a college education cannot be measured in money, but tuition is as high as a kite.
4. As he looked back, the farmer thought of those mornings as cold as ice when the ground was as hard as a rock and when he shook like a leaf as he rose at the crack of dawn.

5. The survivor was <u>weak as a kitten</u> after eight days on the ocean, but in a few days he was <u>fit as a fiddle</u>.
6. The brothers were <u>as different as night and day</u>, but each <u>drank like a fish</u>.
7. The heiress claimed that she wanted to marry <u>a strong, silent type</u>, but she <u>tied the knot</u> with her hairdresser.
8. It was <u>raining cats and dogs</u>, but still she managed to look as <u>pretty as a picture</u>.
9. Surely it is possible to <u>keep abreast of the times</u> without giving up the <u>tried and true</u> values of the past.
10. If you <u>pinch pennies</u> now, you can enjoy <u>the golden years</u> with more <u>peace of mind</u>.

38f Use fresh and effective figures of speech. Avoid mixed and inappropriate figures.   *fig*

Figures of speech compare one thing (usually abstract) with another (usually literal or concrete). Mixed figures associate things that are not logically consistent.

MIXED

> These corporations lashed out with legal loopholes. [You cannot *strike* with a *hole*.]

Inappropriate figures of speech compare one thing with another in a way that violates the mood or the intention.

INAPPROPRIATE

> Shakespeare is the most famous brave in our tribe of English writers. [It is inappropriate and puzzling to compare Shakespeare to an Indian brave and a group of writers to an Indian tribe.]

Use figurative comparisons (of things not literally similar) for vivid explanation and for originality. A simile, a metaphor, or a personification gives you a chance to compare or to explain what you are saying in a different way from the sometimes prosaic method of pure statement, argument, or logic.

METAPHORS (implied comparisons)

Though calm without, the young senator was a volcano within.

Old courthouse records are rotting leaves of the past.

SIMILES (comparisons stated with *like* or *as*)

Though calm without, the young senator was like a volcano within.

Old courthouse records are like rotting leaves of the past.

■ Exercise 17

Explain the flaws in these figures of speech.

1. The new race car flew once around the track and then limped into the pit like a sick horse into its stable.

   Comparing a machine first to a bird and then to a horse is a flawed mixture.

2. The speaker's flamboyant oration began with all the beauty of the song of a canary.

   The volume, pitch, and rhythm of an oration are not embodied well in a sound like the song of a canary.

3. Crickets chirped with the steadiness of the tread of a marching army.

   The pounding of a marching army does not describe the sounds of crickets well.

4. The warm greetings of the students sounded like a horde of apes rushing through a jungle.

   Mannerly human responses are not well expressed in the simile of rushing apes.

5. He nipped the plan in the bud by pouring cold water on all suggestions.

   Nipping (cutting) is not similar to pouring a cold liquid.

■ Exercise 18

Compose and bring to class two fresh and appropriate figures of speech.

■ Exercise 19

Find two figures of speech in your reading, and show how they are mixed or especially appropriate or inappropriate.

Papers

---

39 Writing Papers

A good paper usually evolves through a number of stages, including preliminary thinking (brainstorming), jotting down ideas and writing in a free-flowing way (freewriting), composing a draft (or drafts), revising, and writing the final version.

39a Generate and develop ideas.

Whether your instructor assigns you a topic to write about or expects you to discover your own, you should make a beginning by coming up with ideas and getting them on paper or on the screen of your computer. Some techniques proven effective in generating and developing ideas are discussed below.

Brainstorming

Brainstorming begins with opening up the mind to sudden flurries of ideas. This mental activity is often associated with unpredictable inspiration, but with some practice you can initiate it whenever you wish. Force yourself to open your mind to all possibilities and to make free associations. As ideas occur to you, get them down rapidly on paper or on your computer screen. Later you can go over the results of your brainstorming to determine which ideas to keep and which to reject.

Journal jotting

Many a good idea for a paper has been lost—that is, forgotten—because it was not recorded when it emerged. You cannot depend on even a sharp and reliable memory to retain fleeting ideas. Therefore, develop the habit of **journal jotting**. Some of the world's greatest writers keep a journal and have it available at all times so that they can write down thoughts that can later be developed. Unless you want to make the entries extensive, you need not do so; jot down just enough to make sure you can remember a promising idea at a later time.

Listing

You may find it helpful to **list** various points suggested by your general subject. At this stage you do not need to be concerned with their order or final relevance. In fact, you will probably list more points than you will want to make in your paper. The purpose of listing, and of other prewriting strategies, is to get your thoughts on paper and to stimulate further thinking.

Suppose a student is interested in current advertising and has decided to write a paper on some aspect of this subject. He or she might list the following thoughts:

1. recent advances in ways of presenting products
2. doctors and attorneys who advertise
3. selling by mail
4. news commentators who also deliver commercials
5. opportunities in this field for high-paying positions
6. may be the most competitive of all fields at present
7. government control of advertising—is it a good idea?
8. for successful firms, is advertising really necessary?
9. effects of disagreeable advertisements on sales
10. artistic content of some television advertisements
11. talk shows that allow guests to plug latest motion picture or book
12. deceptiveness in methods of advertising
13. advertisements in newspapers that are made to look like news items
14. appeals for money on religious programs
15. false advertising
16. the high cost of television advertising
17. dangers of disguised advertising—brings on loss of trust
18. ethics—question of whether still important to those who control advertising

19. advertising and mind control
20. subliminal messages

The writer has randomly listed almost everything that came to mind—too much for a paper of average length. The list is a good start, however, in exploring the subject of advertising.

Clustering

Relationships among several ideas are often easier to understand when you project them graphically. In **clustering** you draw a diagram rather than write. Just as graphs and maps are useful alternatives to words, this technique of invention and development can aid in revealing connections among ideas. Begin by writing down your principal subject and circling it. Then as sub-ideas come to you, jot them down, circle them, and draw connecting lines to reveal relationships that you have perceived.

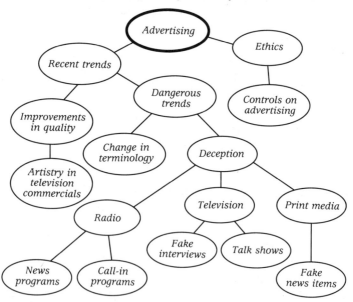

A large blackboard is ideal for this kind of exercise when you are developing a subject, working through a mental puzzle. You can erase frequently and substitute new ideas. If a blackboard is not available, you can achieve much the same effect with pencil and paper. However you do it, clustering enables you to view the subject from different perspectives and to see graphically how the composite parts are related.

Freewriting

Some writers prefer not to begin writing before they have given their subject a good deal of thought and planning; others find that the discovery process can occur during the act of writing. The problem is how to get started. One way is to begin by expressing in words any thoughts that come to mind. This activity, called **freewriting,** takes a writer where whims dictate. At some point a degree of coherence emerges out of fragmentation; related ideas begin to appear.

Freewriting usually does not produce a first draft. It is a preliminary step—the written form of brainstorming. It is a way of discovering what you want to write about as you write, a way of overcoming the inhibitions some people experience when they view a blank sheet of paper or a blank screen.

39b Compose a precise thesis statement.

Ideas emerging from the preliminary processes described above are likely to be somewhat in disarray. They will need to be examined closely, culled, and ordered. This stage of the writing process is necessary if the finished paper is to be coherent and sharply focused, not rambling and pointless.

At some point you must determine precisely what it is you wish to say. Each paper should have a clear center from which all the details and illustrations radiate.

When you pin down the central idea, you can usually express it in a single sentence, called the **thesis statement**. Many instructors require you to compose a specific and concise statement and to include it in your first or second paragraph to develop a good sense of direction.

A thesis statement tells in a few words what your argument is, not what your argument is *about*. Compare the three statements below (see the model paper "Conduct and the Inner Self," pp. 255–257, written in response to Willard Gaylin's essay "What You See Is the Real You," pp. 236–238).

NOT

> The purpose of this paper is to agree basically with Willard Gaylin's essay "What You See Is the Real You." [A description, not a thesis statement.]

NOT

> In dealing with people, the profession of psychoanalysis often gives society the wrong impression about moral responsibility. [A thesis statement, but not a good one because it is not specific enough.]

BUT

> Studying and emphasizing "the inner man," as Willard Gaylin argues in "What You See Is the Real You," has had the unfortunate effect of relieving individuals of moral responsibility for their acts.

39c Organize effectively.

What to place where can be a problem—although sometimes the parts fall naturally into place. You may or may not need an outline, depending on the extent to which the development process has already unfolded in your mind. It is a good idea to set out with some sort of plan, though, even if you require only a brief one. Three of the most common kinds of outlines are described below.

Scratch outline

This is the simplest kind of outline: a list of points to make, in order but without subdivisions. It is a quick way to organize your thoughts and to remind yourself of their order while you are writing. For brief papers, those written in class, and essay examinations, a scratch outline usually suffices. You might use the following points for a scratch outline of the paper entitled "Conduct and the Inner Self."

> Gaylin right in pointing out modern tendency to stress inner self
> importance of accepting moral responsibility for actions
> history proves this
> Gaylin in error in some of his illustrations
> soundness of his basic point

Topic outline

The topic outline is a formal, detailed structure to help you organize your materials. Observe the following conventions:

1. Number the main topics with Roman numerals, the first subheadings with capital letters, and the next with Arabic numbers. If further subheadings are necessary, use a, b, c and (1), (2), (3).

    I. ....
       A. ....
          1. ....
             a. ....
                (1) ....
                (2) ....
             b. ....
          2. ....
       B. ....
    II. ....

2. Use parallel grammatical structures.
3. Write down topics, not sentences.

4. Do not place periods after the topics.
5. Punctuate as in the example that follows.
6. Check to see that your outline covers the subject fully.
7. Use specific topics and subheadings arranged in a logical, meaningful order.
8. Be sure that each level of the outline represents a division of the preceding level and has a smaller scope.
9. Avoid single subheadings. If you have a Roman numeral I, you should also have a II, and so forth.

INCORRECT

    I. Gaylin's essay as a corrective to some current thinking
       A. Nature of faulty thinking on individual responsibility
       ←——————— *B. [Another subheading is needed.]*
    II. Gaylin's argument against the reality of the "inner man"

Following is an example of a topic outline for the paper "Conduct and the Inner Self":

Conduct and the Inner Self

    I. Gaylin's essay as corrective to some current thinking
       A. Nature of faulty thinking about individual responsibility
       B. How study of "inner man" contributes to faulty thinking
    II. Gaylin's argument against the reality of the "inner man"
       A. Purpose of argument
       B. Argument from historical perspective
    III. Problems in Gaylin's presentation
       A. Exaggeration
          1. Praise of a hypocrite
          2. Total denial of the good-heart principle
       B. Contradiction
    IV. Validity of Gaylin's argument despite flaws

Sentence outline

The sentence outline is an extensive form of preparation for writing a paper. More thinking goes into it than into a scratch or topic outline, but the additional effort usually enables you to keep tight control over your writing. The sentence outline follows the same conventions as the topic outline, but the entries are expressed in complete sentences. Place periods after sentences in a sentence outline.

Conduct and the Inner Self

I. Gaylin's essay is a much-needed corrective to faulty thinking today.
   A. Currently there is a tendency to excuse those who act irresponsibly.
   B. Emphasizing the "inner man" tends to relieve individuals of moral responsibility.
II. Gaylin's argument against the reality of the "inner man" is ingenious and sound.
   A. The purpose of the argument is to stress that no aspect of our being is more important than our conduct.
   B. History has proven Gaylin correct in his insistence on personal responsibility.
III. Some aspects of Gaylin's argument present problems.
   A. He exaggerates to make a point.
      1. Praise of the old man who lived a life of hypocrisy is unconvincing.
      2. Gaylin's denial that a heart of gold can lie beneath bad behavior contradicts experience.
   B. Gaylin contradicts himself when he argues against the existence of an inner self and then speaks as if there were such a thing.
IV. Though flawed, Gaylin's essay makes a valid and timely point.

39d   Adapt your writing to your audience.

Writing a paper is an act of communication, not merely an exercise or an opportunity for self-expression. Prepare yourself to reach your audience effectively. Consider the following questions. How much information do members of the audience have on the subject you are dealing with? What is their average level of education? Are they likely to have strong opinions one way or the other on the topic? What lines of work or study are they engaged in as a rule? Do they fall into a definite age group? Writers who ignore such important considerations are inviting rejection. Mark Twain learned the necessity of carefully matching writing with audience when he composed and delivered a humorous but rather crude series of caricatures to an audience of distinguished and highly dignified New England writers whom he greatly admired and did not wish to insult. The performance was received with stony silence by the offended audience, and Mark Twain regretted his misjudgment for life.

Although your instructor may ask you to write for specialized audiences, usually papers are composed with *general readers* in mind: educated adults who do not have highly specialized vocabularies and who are interested in a wide variety of general subjects. To capture their attention, you must be intelligent, interesting, rational, and clear. To keep their attention and win their respect, you must be fair and remain sensitive to their feelings. For example, you should be constantly aware that your audience consists of both men *and* women and avoid using generic masculine pronouns (*he* for *he or she*) and terms like *man* and *mankind* (for *humankind*). Notice that Willard Gaylin uses terms of masculine gender ("inner man") in his thought-provoking essay on pp. 236–238 and thus runs the risk of offending and alienating a segment of his audience. Study the model paper (pp. 255–257) written as a response to Gaylin's essay, and observe how the author of the paper wisely avoids language that could be interpreted as sexist (see also **8d** and p. 461).

Resist the temptation to write solely for an individual professor. Frustrated students who complain that they do not know what their professors want in a paper can often avoid this needless exercise in trial and error by concentrating on a general audience. If you compose with only your professor in mind, you may write somewhat over your head or employ obscure or flowery language.

Audience awareness helps you to become a better judge of your own writing and consequently to revise more successfully. By learning to read your writing as your audience will, you can communicate with greater clarity and grace.

■ Exercise 1

Think of an experience in which you felt that you were wronged or in which you wronged someone: something involving a store, an automobile repair shop, a restaurant, an insurance company, or the like. Then write about this experience three times, with three audiences in mind: (1) close friends; (2) managers, supervisors, or owners of the business; (3) readers of a large local newspaper. All three writings may be in the form of letters.

39e Sustain an appropriate tone.

Tone derives from attitude. If you are displeased, you can show that attitude through your choice of words. If you are amused and wish to pass that attitude on to your audience, you can do so by creating a humorous tone. Enthusiasm, urgency, fear, objectivity, displeasure, playfulness, skepticism—all can be conveyed effectively, but first you have to determine precisely what attitude you wish to communicate. Trouble occurs when a writer is uncertain what tone to take and suddenly switches unthinkingly from one attitude to another, as in the following example:

> Art critics have the power to make or break young painters. Their word is especially influential among collectors and wealthy investors who deal in art. If a few of these better-known critics praise a given artist, he or she can rise almost overnight from struggling anonymity to wealthy repute. Conversely, all it takes is a negative opinion from them to doom an aspiring artist. Do these critics exercise their *shift* enormous power responsibly? They care only for themselves. They *in* are shameless frauds who invent nonsense about art and pass it off *tone* as the truth. They are arrogant and effete impostors who deserve to be whipped.

At the beginning of the passage, emotions are under control and the tone is coolly objective. Then the tone changes abruptly as an angry outburst destroys what could otherwise be a convincing argument. If the writer had decided ahead of time to make indignation the prevailing tone of the essay, he or she could have written indignantly but thoughtfully and effectively. As it is, the writing is ineffectual because the sudden shift in tone confuses readers.

Whatever tone you choose, sustain it carefully; do not shift back and forth from one attitude to another. In general, avoid flippancy and sarcasm. Flippancy in writing is often the mark of immaturity. Sarcasm is difficult to sustain without creating hostility.

39f Compose a draft before writing the final version.

A rough draft is a necessary step toward a finished paper. Writers who go directly from an outline to what they consider the final copy usually produce a less effective paper than they could otherwise. Do not worry about how you are expressing your ideas in your draft; simply get them down. Then you can make improvements.

The rough draft on pp. 239–243 emerged from an assignment in which the student was required to read a specific essay and then to respond to it in a paper of about five hundred words. First read the essay on which the assignment was based, "What You See Is the Real You" by Willard Gaylin. Then read the student's rough draft.

What You See Is the Real You
Willard Gaylin

It was, I believe, the distinguished Nebraska financier Father Edward J. Flanagan who professed to having "never met a bad boy." Having, myself, met a remarkable number of bad boys, it might seem that either our experiences were drastically different or we were using the word "bad" differently. I suspect neither is true, but rather that the Father was appraising the "inner man," while I, in fact, do not acknowledge the existence of inner people.

Since we psychoanalysts have unwittingly contributed to this confusion, let one, at least, attempt a small rectifying effort. Psychoanalytic data—which should be viewed as supplementary information—is, unfortunately, often viewed as alternative (and superior) explanation. This has led to the prevalent tendency to think of the "inner" man as the real man and the outer man as an illusion or pretender.

While psychoanalysis supplies us with an incredibly useful tool for explaining the motives and purposes underlying human behavior, most of this has little bearing on the moral nature of that behavior.

Like roentgenology, psychoanalysis is a fascinating, but relatively new, means of illuminating the person. But few of us are prepared to substitute an X-ray of Grandfather's head for the portrait that hangs in the parlor. The inside of the man represents another view, not a truer one. A man may not always be what he appears to be, but what he appears to be is always a significant part of what he is. A man is the sum total of *all* his behavior. To probe for unconscious determinants of behavior and then define him in their terms exclusively, ignoring his overt behavior altogether, is a greater distortion than ignoring the unconscious completely.

Kurt Vonnegut has said, "You are what you pretend to be," which is simply another way of saying, you are what we (all of us) perceive you to be, not what you think you are.

Consider for a moment the case of the ninety-year-old man on his deathbed (surely the Talmud must deal with this?) joyous and relieved over the success of his deception. For ninety years he has affected courtesy, kindness, and generosity—suppressing all the malice he knew was within him while he calculatedly and artificially substituted grace and charity. All his life he had been fooling the world into believing he was a good man. This "evil" man will, I predict, be welcomed into the Kingdom of Heaven.

Similarly, I will not be told that the young man who earns his pocket money by mugging old ladies is "really" a good boy. Even my generous and expansive definition of goodness will not accommodate that particular form of self-advancement.

It does not count that beneath the rough exterior he has a heart—or, for that matter, an entire innards—of purest gold, locked away from human perception. You are for the most part what you seem to be, not what you would wish to be, nor, indeed, what you believe yourself to be.

Spare me, therefore, your good intentions, your inner sensitivities, your unarticulated and unexpressed love. And spare me also those tedious psychohistories which—by exposing the goodness inside the bad man, and the evil in the good—invariably establish a vulgar and perverse egalitarianism, as if the arrangement of what is outside and what inside makes no moral difference.

Saint Francis may, in his unconscious, indeed have been compensating for, and denying, destructive, unconscious Oedipal impulses identical to those which Attila projected and acted on. But the similarity of the unconscious constellations in the two men matters precious little, if it does not distinguish between them.

I do not care to learn that Hitler's heart was in the right place. A knowledge of the unconscious life of the man may be an adjunct to understanding his behavior. It is *not* a substitute for his behavior in describing him.

The inner man is a fantasy. If it helps you to identify with one, by all means, do so; preserve it, cherish it, embrace it, but do not present it to others for evaluation or consideration, for excuse or exculpation, or, for that matter, for punishment or disapproval.

Like any fantasy, it serves your purpose alone. It has no standing in the real world which we share with each other. Those character traits, those attitudes, that behavior—that strange and alien stuff sticking out all over you—*that's the real you!*

Rough draft

An Analysis of "What You See Is the Real You"

There are some things that I don't agree with in it but on the whole, Willard Gaylin's essay, "What You See Is the Real You", is a much-needed corrective to much falty thinking today. First of all, as Gaylin himself indicates to the reader, he is a psychoanalyst who is not in full agreement with the curent practice of his perfession. The reader sees with some surprise that he apparently believes that in his own field as well as in other fields, there is a tendency to excuse those who act irresponsibly or even criminally, because, as the theory goes, deep down they may be decent and caring people. Gaylin informs the reader that psychoanalysis is good but studying and emphasizing the "inner man," to use Gaylin's own term, has had the effect of relieving individuals of the moral responsibility of their actions.

I think that he wrote this article, because he wanted to tell readers that despite all this bull about criminals and other undesireables being that way

because of the way they were brought up and that be—
neath their badness is something good just waiting to
be brought out, people are really responsible for
their actions. This is what this essay really says to
me. You can't put the blame on environment or bad
treatment or neglect or poverty or abuse or anything
else. You are responsible for your actions, my
friend. Period.

One of the most engenuous things about the essay
is the way the author gets the reader to think this
way about the need of individual responsibility. The
reader gets so fascinated with the idea of the "inner
man" that he/she is subtly brought around to the au—
thor's way of thinking. I have noticed this technique
in several of the essays we have studied this term.
An author will manipulate the reader by arguing along
one line with the purpose of accomplishing something
else. The essay by Swift that we studied is a good
example of this. But anyway, the reader can see that
Gaylin is engenuous in his use of the argument against
the reality of "inner people," because in actuality he
is not so much concerned with debating this psycho—

logical and philosophical subject as he is with finding a way to stress to the reader that it is both
inaccurate and counterproductive to blame unacceptable
behavior on one's environment. Or on any other factors for that matter. Gaylin denies the existence of
an inner self in order to emphasize to the reader that
there is no aspect of our being more basic and fundamental, more important, than our conduct. Gaylin
urges that we integrate our sense of self with our
behavior and think of them as one by doing so, readers
will realize that they are responsible for their own
behavior. History, the reader tends to agree, has
certainly proven Gaylin correct in his insistence on
personal responsibility. The reader can just imagine
the chaos that would have resulted down through the
centuries if budding civilizations had taken a kind of
no-fault reasoning. Failing to expect people to abide
by laws and become productive citizens, forgiving them
for all sorts of antisocial conduct, because of professed good intentions or disadvantaged upbringing.

   I believe that some aspects of Gaylin's argument,
however, do present problems. He goes too far. In an

effort to make his point, he exaggerates and even en-
volves himself in contradictions. I think the old man
is a good example of this. Gaylin's praise amounts to
praise of hypocrisy. In addition, Gaylin goes too far
in denying that a heart of gold may belie a deceptive
exterior. Readers have probably known a person who
most of the time acts selfishly but who on occasion
behaves in such a way to suggest that there is real
goodness in his nature. I believe Gaylin seems to
question that there ever could be such a person and in
doing so he fails in this part of his essay to be
convincing.

Although the essay may be logically flawed, it
appeals to the reader nevertheless as making a valit
and powerful point that should not be obscured by an
exaggeration or a contradiction here and there. We
act as we do because of what we are, and we must take
full responsibility for our acts. It is not who or
what we think we are that counts but how we behave.

A rather serious question, however, arises to the
reader from his saying early in the essay that he does
not acknowledge the existence of inner people but then

he procedes as if he does, for he compares the "inner
man" to an X-ray, real but distinctively different
from a portrait. I would like to ask the question, if
the inner self is a "fantasy," as he says later, then
how can it be X-rayed?

■ Exercise 2

*Make two lists, one that notes what is good in the above draft and
one that points out problems and errors (weaknesses that you
would work on if you were revising the paper).*

39g Analyze your draft.

Develop a critical eye for examining your own writing. Stand apart
from it, and imagine that it belongs to someone else. Make a list
of the major weaknesses in your draft, the problems to work on
in the revision.

Follow the process of revision in section **39h** to see how the
student shaped the draft on pp. 239–243 into a good paper.

39h Revise the draft.

Generally it is a good idea to take care of the larger matters
(sometimes called **global revisions**) first. When you revise globally,
you delete irrelevant materials, add necessary passages, and move
sentences or whole sections around. Note these larger revisions
in the draft below. The second paragraph has been deleted because
it is badly written and repeats points suggested elsewhere. The
first several lines of the third paragraph have been cut because

they contain material that is not relevant to the subject. Words have been added in the fourth paragraph to make clear the reference to the "old man." The final paragraph is anticlimactic, but it will work well if moved to a new position, as indicated.

Draft with global revisions

An Analysis of "What You See Is the Real You"

There are some things that I don't agree with in it but on the whole, Willard Gaylin's essay, "What You See Is the Real You", is a much-needed corrective to much falty thinking today. First of all, as Gaylin himself indicates to the reader, he is a psychoanalyst who is not in full agreement with the curent practice of his perfession. The reader sees with some surprise that he apparently believes that in his own field as well as in other fields, there is a tendency to excuse those who act irresponsibly or even criminally, because, as the theory goes, deep down they may be decent and caring people. Gaylin informs the reader that psychoanalysis is good but studying and emphasizing the "inner man," to use Gaylin's own term, has had the effect of relieving individuals of the moral responsibility of their actions.

I think that he wrote this article, because he wanted to tell readers that despite all this bull about criminals and other undesireables being that way because of the way they were brought up and that beneath their badness is something good just waiting to be brought out, people are really responsible for their actions. This is what this essay really says to me. You can't put the blame on environment or bad treatment or neglect or poverty or abuse or anything else. You are responsible for your actions, my friend. Period.

One of the most engenuous things about the essay is the way the author gets the reader to think this way about the need of individual responsibility. The reader gets so fascinated with the idea of the "inner man" that he/she is subtly brought around to the author's way of thinking. I have noticed this technique in several of the essays we have studied this term. An author will manipulate the reader by arguing along one line with the purpose of accomplishing something else. The essay by Swift that we studied is a good example of this. But anyway, the reader can see that

delete

¶ Gaylin is engenuous in his use of the argument against the reality of "inner people," because in actuality he is not so much concerned with debating this psychological and philosophical subject as he is with finding a way to stress to the reader that it is both inaccurate and counterproductive to blame unacceptable behavior on one's environment. Or on any other factors for that matter. Gaylin denies the existence of an inner self in order to emphasize to the reader that there is no aspect of our being more basic and fundamental, more important, than our conduct. Gaylin urges that we integrate our sense of self with our behavior and think of them as one by doing so, readers will realize that they are responsible for their own behavior. History, the reader tends to agree, has certainly proven Gaylin correct in his insistence on personal responsibility. The reader can just imagine the chaos that would have resulted down through the centuries if budding civilizations had taken a kind of no-fault reasoning. Failing to expect people to abide by laws and become productive citizens, forgiving them

for all sorts of antisocial conduct, because of professed good intentions or disadvantaged upbringing.

I believe that some aspects of Gaylin's argument,
however, do present problems. He goes too far. In an
effort to make his point, he exaggerates and even en-
volves himself in contradictions. ~~I think the old man
is a good example of this.~~ Gaylin's praise ✓amounts to
praise of hypocrisy. In addition, Gaylin goes too far
in denying that a heart of gold may belie a deceptive
exterior. Readers have probably known a person who
most of the time acts selfishly but who on occasion
behaves in such a way to suggest that there is real
goodness in his nature. I believe Gaylin seems to
question that there ever could be such a person and in
doing so he fails in this part of his essay to be
convincing.
Put last paragraph here.
Although the essay may be logically flawed, it
appeals to the reader nevertheless as making a valit
and powerful point that should not be obscured by an
exaggeration or a contradiction here and there. We
act as we do because of what we are, and we must take

*His apparent admiration for the old man
who had lived a life of deception — thinking
one way and acting another —*

full responsibility for our acts. It is not who or
what we think we are that counts but how we behave.

> A rather serious question, however, arises to the
> reader from his saying early in the essay that he does
> not acknowledge the existence of inner people but then
> he procedes as if he does, for he compares the "inner
> man" to an X-ray, real but distinctively different
> from a portrait. I would like to ask the question, if
> the inner self is a "fantasy," as he says later, then
> how can it be X-rayed?

Move this paragraph to previous page.

After global revisions, you can turn to **local** (or **sentence-
level**) **revisions**—that is, the correcting of errors in spelling,
punctuation, and mechanics, and the clearing up of problems
with wordiness, repetition, diction, and so forth. Study the corrections that have been made in the following draft. A new title
has been supplied. (Avoid expressions like "A Reading of" and
"An Analysis of" in titles. See **42b**.) The style has been tightened
throughout the essay by deleting useless references to "the reader"
(see **42f**) and expressions using the first-person pronoun *I*. Note
other changes to correct repetition and redundancy, misspellings,
and misuse of commas.

Draft with local revisions

Conduct and the Inner Self
~~An Analysis of "What You See Is the Real You"~~

~~There are some things that I don't agree with in~~
~~it but~~ on the whole, Willard Gaylin's essay, "What You
See Is the Real You", is a much-needed corrective to
faulty
~~much~~ falty thinking today. ~~First of all,~~ As Gaylin
himself indicates ~~to the reader~~, he is a psychoanalyst
                                        current
who is not in full agreement with the curent practice
         profession
of his perfession. ~~The reader sees with some surprise~~
         H
~~that~~ he apparently believes that in his own field as
                      s
well as in other fields, there is a tendency to excuse
those who act irresponsibly or even criminally, because, as the theory goes, deep down they may be decent and caring people. ~~Gaylin informs the reader~~
                                S
~~that psychoanalysis is good but~~ studying and emphasizing the "inner man," to use Gaylin's own term, has
had the effect of relieving individuals of the moral
responsibility of their actions.

Gaylin is ~~engenuous~~ *ingenious* in his use of the argument against the reality of "inner people~~;~~." ~~because~~ *In* actuality he is not so much concerned with debating this psychological and philosophical subject as he is with finding a way to stress ~~to the reader~~ that it is ~~both~~ inaccurate and counterproductive to blame unacceptable behavior on one's environment~~,~~ *or* on any other factors~~, for that matter.~~ Gaylin denies the existence of an inner self in order to emphasize ~~to the reader~~ that there is no aspect of our being more basic ~~and funda-mental~~, more important, than our conduct. ~~Gaylin~~ *He* urges that we integrate our sense of self with our behavior and think of them as one. *B*~~by~~ doing so, ~~readers~~ *people* will realize that they are responsible for their own behavior. ¶History~~, the reader tends to agree,~~ has certainly proven Gaylin correct in his insistence on personal responsibility. ~~The reader~~ *One* can ~~just~~ imagine the chaos that would have resulted down through the centuries if budding civilizations had ~~taken~~ practiced a kind of

no-fault reasoning~~,~~. ~~F~~failing to expect people to abide

by laws and become productive citizens, forgiving them

for ~~all sorts of~~ antisocial conduct~~,~~ because of pro-

fessed good intentions or disadvantaged upbringing.

~~I believe that~~ Some aspects of Gaylin's argument,

however, do present problems. ~~He goes too far.~~ In an

effort to make his point, he exaggerates and even ~~en-volves~~ involves himself in contradictions. His apparent admi-

ration for the old man who had lived a life of decep-

tion--thinking one way and acting another--amounts to

praise of hypocrisy. In addition, Gaylin goes too far

in denying that a heart of gold may ~~belie~~ underlie a deceptive

exterior. Almost everyone has ~~Readers have probably~~ known a person who

most of the time acts selfishly but who on occasion

behaves in such a way to suggest that there is real

goodness in his or her nature. ~~I believe~~ Gaylin seems to

question that there ever could be such a person, and in

doing so he fails in this part of his essay to be

convincing.

A ~~rather~~ *Somewhat more* serious question, ~~however,~~ arises ~~to the reader~~ from his saying early in the essay that he does

"
ˆnot acknowledge the existence of inner people" but then

proceeding
~~he procedes~~ as if he does, for he compares the "inner

man" to an X—ray, real but distinctively different

from a portrait. ~~I would like to ask the question,~~ ˆIf
I

the inner self is a "fantasy," as he says later, then

how can it be X—rayed?

Although the essay may be logically flawed, it

makes
~~appeals to the reader nevertheless as making~~ a ~~valit~~ *valid*

and powerful point that should not be obscured by an

exaggeration or a contradiction here and there. We

act as we do because of what we are, and we must take

full responsibility for our acts. It is not who or

what we think we are that counts but how we behave.

Before submitting your paper, read it over two or three times, at least once aloud. Listen for annoying repetitions. Compare the finished copy of the paper "Conduct and the Inner Self" on pp. 255–257 to the draft on pp. 239–243.

You will find the following checklist helpful during the process of writing and revising as well as just before turning in your paper.

Checklist

Title

The title should accurately suggest the contents of the paper.

It should attract interest without being excessively novel or clever.

It should not be too long.

NOTE: Do not underline the title of your own paper, and do not put quotation marks around it.

Introduction

The introduction should be independent of the title; no pronoun or noun in the opening sentence should depend for meaning on the title.

It should catch the reader's attention.

It should properly establish the tone of the paper as serious, humorous, ironic, or otherwise.

It should include a thesis statement that declares the subject and the purpose directly but at the same time avoids worn patterns like "It is the purpose of this paper to . . ."

Body

The materials should develop the thesis statement.

The materials should be arranged in logical sequence.

Strong topic sentences (see **41b**) should clearly indicate the direction in which the paper is moving and the relevance of the paragraphs to the thesis statement.

Technical terms should be explained.

Paragraphs should not be choppy.

Adequate space should be devoted to main ideas; minor ideas should be subordinated.

Concrete details should be used appropriately; insignificant details should be omitted.

Transitions

The connections between sentences and those between paragraphs should be shown by appropriate linking words and by repetition of parallel phrases and structures (see **41h**).

Conclusion

The conclusion should usually contain a final statement of the underlying idea, an overview of what the paper has demonstrated.

The conclusion may require a separate paragraph; but if the paper has reached significant conclusions all along, such a paragraph is not necessary for its own sake.

The conclusion should not merely restate the introduction.

Proofreading

Allow some time, if possible at least one day, between the last draft of the paper and the final, finished copy. Then you can examine the paper objectively for wordiness, repetition, incorrect diction, misspellings, faulty punctuation, choppy sentences, vague sentences, lack of transitions, and careless errors.

39i Model paper

Conduct and the Inner Self

On the whole, Willard Gaylin's essay "What You See Is the Real You" is a much-needed corrective to faulty thinking today. As Gaylin himself indicates, he is a psychoanalyst who is not in full agreement with the current practice of his profession. He apparently believes that in his own field as well as in others, there is a tendency to excuse those who act irresponsibly or even criminally because, as the theory goes, deep down they may be decent and caring people. Studying and emphasizing the "inner man," to use Gaylin's term, has had the effect of relieving individuals of the moral responsibility of their actions.

Gaylin is ingenious in his use of the argument against the reality of "inner people." In actuality, he is not so much concerned with debating this psychological and philosophical subject as he is with finding a way to stress that it is inaccurate and counterproductive to blame unacceptable behavior on one's environment or any other factors. He denies the existence of an inner self in order to emphasize that there is no aspect of our being more basic or important than our conduct. He urges that we integrate our sense of self with our behavior and think of them as one. By doing so, people will realize that they are responsible for their own behavior.

History has certainly proven Gaylin correct in his insistence on personal responsibility. One can imagine the chaos that would have resulted down through the centuries if budding civilizations had practiced a kind of no-fault reasoning, failing to expect people to abide by laws and become productive citizens, forgiving them for anti-social conduct because of professed good intentions or disadvantaged upbringing.

Some aspects of Gaylin's argument, however, do present problems. In an effort to make his point, he exaggerates and even involves himself in contradictions. His apparent admiration for the old man who had lived a life of deception—thinking one way and acting another—amounts to praise of hypocrisy. In addition, Gaylin goes too far in denying that a heart of gold may underlie a deceptive exterior. Almost everyone has known a person who most of the time acts selfishly but who on occasion behaves in such a way to suggest that there is real goodness in his or her nature. Gaylin seems to question that there ever could be such a person, and in doing so he fails in this part of his essay to be convincing.

A somewhat more serious question arises from his saying early in the essay that he does "not acknowledge the existence of inner people" but then proceeding as if he does, for he compares the "inner man" to an X-ray, real but distinctively different from a portrait. If the in- ner self is a "fantasy," as he says later, then how can it be X-rayed?

Although the essay may be logically flawed, it makes a valid and powerful point that should not be obscured by an exaggeration or contradiction here and there. We act as we do because of what we are, and we must take full responsibility for our acts. It is not who or what we think we are that counts but how we behave.

39j Learn from your mistakes.

The process of composition is not complete until you have carefully examined your paper after it is returned to you, looked up your errors in this book and in a dictionary so that you understand how to correct them, and revised along the lines suggested by your instructor. Learn from your mistakes so that the same problems do not turn up in future papers.

■ Exercise 3

Write a critique (500–600 words) of either Willard Gaylin's essay "What You See Is the Real You" (pp. 236–238) or the model paper "Conduct and the Inner Self" (pp. 255–257). Evaluate the strengths, and point out any weaknesses in the author's argument.

39k Composing and revising on a word processor.

Most of the writing techniques discussed in this chapter—for example, listing and freewriting—can be effectively practiced on a personal computer with a word-processing program. Many writers find it easier to initiate the flow of ideas on a word processor than on a typewriter or with pen and paper. Thoughts can be recorded on the screen and discarded or retained quickly and easily. The popularity of word processors is based not merely on their ability to save time; they enable users to engage in the process of creating without seriously interrupting the flow of mental current.

If you are used to writing drafts of papers in longhand or on a typewriter, you probably have a built-in resistance to throwing away sheets of paper and beginning all over again, recopying, retyping, and laboriously correcting mistakes. The awareness of inconvenience often acts as a psychological barrier to thorough revisions of drafts. With a word processor you can insert words, sentences, and paragraphs, make minor or extensive changes in wording, delete passages, check spelling, find good synonyms, and revise in other ways with an ease undreamed of several years ago. Use these ways of making changes, and cultivate the habit of thorough revision, the best habit a writer can develop.

Though by no means complete, the following list will furnish some helpful suggestions about writing on a word processor.

Save.

Use the **Save** command to protect yourself against losing what you have written. Some mistake—hitting the wrong keys, disconnecting or cutting off the machine—a power surge, or other mishap could erase hours of work. Different versions of a paper can be saved under different titles and then profitably compared during the process of revision.

Print out.

Regularly print out copies of what you are working on. Seeing on paper what you have been viewing on the screen gives you another perspective and makes it easier to spot errors. In addition, if you print out several versions of a document, you may find in an early draft valuable material that you can use, material that would be lost if you simply made all changes on the computer without periodically printing out.

Back up.

Computers do break down; therefore, it is good insurance against losing everything you have written to use a back-up program if you have one, or simply to copy material that you have on the hard disk of your computer onto a diskette (floppy disk) to be stored separately.

Explore.

Most current word-processing programs include a fascinating variety of functions. Learn as much as you can as early as you can about the software you are using. Being thoroughly familiar with the particular computer you are using and with its word-processing program will save you much time later on and increase your effectiveness and your pleasure.

Work comfortably.

Prepare a comfortable working place with easy access to all the materials you will need.

Take frequent breaks.

Sitting at a computer for hours without rest can cause problems with your eyes and neck and make you a less efficient writer.

Replace ribbons regularly.

It is easy to put off changing the ribbon on your printer, but turning in a paper with light type is inconsiderate and sloppy.

Store disks carefully.

Preserve what you have placed on floppy disks by keeping them where they will not be exposed to heat, dampness, or a magnetic field.

Accurate
Thinking and
Writing

40 Logic and Accuracy *log*

Logic and accuracy are indispensable for a sound and convincing argument. Do not talk or write in a way that misleads or gives false information or conclusions. Present facts without error, and reason logically. Admit uncertainties, and take into account valid opposing views or evidence.

Logical argument is either **inductive** or **deductive**.

Inductive method

An inductive argument cites information derived from experiments, examples, cases, related facts, observations, statistics, and so forth to prove a certain conclusion or to establish a principle. This is the scientific method of reasoning, used in many kinds of writing. Inductive argument presents a hypothesis and then cites concrete and specific evidence to prove the reliability of the hypothesis. Trial lawyers often use the inductive method to present cases. From the quality of the evidence the jury determines which argument is likely to reflect the truth just as the audience measures the effectiveness of evidence in an essay employing the inductive method.

Deductive method

A deductive argument derives one conclusion from another; the evidence tends to be other, related conclusions. You would be arguing deductively if you stated (1) that exercise is generally beneficial, (2) that walking is a form of exercise, and (3) that walking is therefore beneficial. To be convincing and logically sound, the statements or premises on which the argument is based must be true. If you said (1) that all colleges are excellent, (2) that Osmosis State is a college, and (3) that Osmosis State is therefore excellent, your argument would be faulty. Obviously you cannot argue forcefully that Osmosis State is excellent because *all* colleges are excellent when it is evident that they are not. In the deductive method, if the truth of the premises on which the conclusion is based is not self-evident, the argument collapses.

Both inductive and deductive argument must stand up to questioning. Of an inductive argument one asks whether the facts are true, whether the exceptions have been noted, whether the selection of materials is representative, whether the conclusions are truly and accurately drawn from the data, and whether the conclusions are stated precisely or exaggerated. Of a deductive argument one asks whether the given principle is impartial truth or mere personal opinion, whether it is applied to materials relevantly, whether the conclusion is accurate according to the principle, and whether exceptions have been noted.

With good motives and bad, with honesty and with deceit, different thinkers reach different conclusions derived from the same data or from the same principles. Learn to discriminate between sound and illogical reasoning. Indeed, when you write, you must constantly question your own reasoning. Watch for errors in thinking—called **logical fallacies**.

40a Use accurate information. Check the facts.

Facts can be demonstrated. They form the bases of judgments. Distinguish carefully between the facts and the judgments derived from them, and then explain how one comes from the other.

Errors, whether of fact, ignorance, or dishonesty, make the reader suspicious and lead to distrust. An otherwise compelling argument crumbles when just one or two facts are shown to be wrong. The following statements contain factual errors.

MISINFORMATION PRESENTED AS FACT

> Columbus was indisputably the first European to step on the North American continent.

FACT

> Some historians assert that other Europeans, especially Vikings, came to America before Columbus. Some historians believe Columbus never set foot on the North American mainland.

MISINFORMATION PRESENTED AS FACT

> Hot water will freeze faster than cold water when placed in a freezer.

FACT

> Physicists state that cold water freezes faster. (However, water that has been heated and has cooled will freeze more quickly than water that has not been heated.)

40b Use reliable authorities.

Not all so-called or self-proclaimed specialists are reliable. Do not accept everything in print as authoritative. Consider the following criteria in evaluating authorities:

1. When was the work published? (An old publication may contain superseded information.)
2. Who published the work? (University presses and well-established publishing houses employ informed consultants, whereas others may not.)
3. Does the work have a reputation for reliability? (For example, how has it been evaluated by other authorities in annotated bibliographies and in reviews?)
4. Is the presumed authority writing about his or her own field? (A noted physician may not be an expert on economic theory.)
5. Are the language and the tone reasonable, or does the authority attempt to persuade by using ornate rhetoric or slanted words?
6. Does the authority show objectivity by admitting the existence of facts that seem contradictory?
7. Does the authority distinguish fact from opinion?

40c Avoid sweeping generalizations. Allow for exceptions.

Do not make statements about *everyone* or *everything* or *all* when you should refer to *some* or *part*. Do not be too inclusive. Such writing can be naive or deceitful. Do not convey wrong information about exceptions to a general principle. Be aware of information and opinions that seem to refute or qualify your generalizations. Deal with them fully and honestly. You can actually strengthen your argument by taking opposing opinions and exceptions into consideration.

SWEEPING GENERALIZATION
    Poor people do not get fair trials in courts.

MORE CONVINCING
    Poor people often do not get fair trials in courts, in many cases because they cannot hire good attorneys.

SWEEPING GENERALIZATION
    Russian athletes are the best in the world.

MORE CONVINCING
    Russian athletes have enjoyed an extraordinary success in world competition.

CAUTION: You cannot justify exaggerations and sweeping generalizations by adding a phrase such as "in my opinion." Overstatement to make a point may irritate, arouse doubt, or create disbelief. The temptation to exaggerate is natural, but moderation (or even understatement) convinces where brashness and arrogance alienate. Generalizations about nationalities and races are often pernicious.

40d Do not reason in a circle.

Reasoning in a circle, often called **begging the question**, begins with an assumption—an idea that requires proof—and then asserts that principle without ever offering proof. It states the same thing twice.

> Universal education is necessary because everyone ought to have an education.

> Some brokers are ambitious simply because they wish to succeed.

40e Avoid false comparisons.

Comparisons are highly useful in a convincing argument. Experienced writers know that a brief but fresh analogy can sometimes be more persuasive than a much longer direct explanation.

Deceptive arguments, however, sometimes try to profit from the effectiveness of analogies by making false comparisons—that is, by comparing two things that have only surface similarities.

FALSE ANALOGY

> This college is much like a Chicago slaughterhouse: neither the freshmen nor the cattle know what is in store for them, and they both get chopped up.

40f Stick to the point.

An argument should be pointed. Do not attempt to be convincing by including irrelevant comments or by shifting the focus to a different issue. The **logical fallacies** discussed below stray from the point and throw up a smoke screen.

Red herring

The term *red herring* originated in the practice of diverting hounds from the scent by moving a fish—a smoked (or red) herring—across their path. The following argument in favor of building a highway tries to win a point by bringing up another subject—crime—that is a separate issue, a red herring.

> Certain neighbors in our city are currently arguing against the construction of a proposed expressway. This highway should be built. Are there not more basic issues, such as crime, that these highway opponents should be addressing?

Argument ad hominem

The Latin term *ad hominem* (literally, "to the man") refers to the fallacy of shifting from the real matter in question to a personal attack, diverting attention from the issue to one's opponent.

> Vote against Senator Wong's bill to provide subsistence to the poor. He has little true interest in the needy; he is a millionnaire who spends much of his time on his plush yacht with his rich friends.

40g   Do not substitute appeals to egotism or prejudice for appeals to reason.

Appeals to egotism or prejudice ignore sound reasoning and attempt to convince through other means.

Name-calling

Simply categorizing what one does not like and giving it a name or label (*fascism, communism, racism, elitism, cultism,* and so forth) does not make a sound argument but appeals to preconceived notions.

> Any investigation into the private life of an elected official is nothing less than blatant McCarthyism.

Flattery

Trying to persuade through praise rather than reason is flattery. The political candidate who tells the people that he knows they will vote for him because of their good character is indulging in this emotional appeal.

Snob appeal

Snob appeal finds its power in the human desire to be a part of the in-crowd. It asserts that one should adopt a certain view because all the better people do.

> The best educated and most progressive readers today will no doubt agree with me that British novels are far superior to any others.

Mass appeal

Related to snob appeal, mass appeal is the **bandwagon** approach to persuasion. It attempts to enlist support of a certain view by asserting that everyone else is supporting it.

> The petition will be circulated for only one more day. People are standing in line to sign it; thousands have already done so; add your name to it while you have the chance.

40h Draw accurate conclusions about cause and effect.

Exact causes of conditions and events are often difficult to determine, just as the effects of an action or a circumstance are. Avoid the two logical fallacies discussed below.

Post hoc, ergo propter hoc

The Latin phrase *post hoc, ergo propter hoc* means "after this, therefore because of this." However, a condition that precedes another is not necessarily the cause of it.

Since Beethoven's greatest music was composed after he lost his hearing, we can assume that deafness was the key to his greatest achievement.

Non sequitur

Non sequitur is Latin for "it does not follow." It does not necessarily follow that one condition in a relationship is logically the effect of another.

Young people who are old enough to drive a car are old enough to raise a family.

40i Avoid the *either . . . or* fallacy (also called false dilemma).

Allow for adequate possibilities. When more than two choices exist, do not illogically assert that there are only two.

Either coauthors must do all their writing together, or they will not be able to write their book.

Children either do very well in school, or they remain illiterate.

40j Avoid illogicalities created by careless wording.

Because of a careless choice of words, what was clear in your mind can be confusing and illogical when it is expressed.

INCORRECT

I was the last of my roommates to arrive home. [Logically, the writer is a roommate but not, as the sentence implies, his or her *own* roommate.]

CORRECT

I arrived home after all my roommates.

Frequently confusion and unintentional humor result when a writer who may be thinking logically does not compose logically:

> When I was barely seventeen, I became engaged. My father was happy and sad at the same time. My mother was just the opposite. [How can one be "the opposite" of "happy and sad at the same time"? What the writer probably means is that her father had mixed feelings about her engagement but her mother was strongly opposed to or strongly approved of it.]

Go over your writing carefully to be sure that you are stating your thoughts logically.

■ Exercise 1

Point out and describe the inaccuracies and illogicalities that you find in the following essay.

The March of Progress

false analogy
[Ensuring progress is a lot like making pickles.

It is pretty sour business sometimes, but the results

are juicy.] On the current American scene are many who

are against cutting the forests for timber, against

using animals for food and animal pelts for garments,
                                                    sweeping
against the use of atomic energy, and so on. [Those
generalization
who are against these things are against progress.]
                    name calling
Most of them are[misfits with psychological problems

who are trying to get attention.]

We cannot turn back the clock. As the population

of the world expands, so must our ideas and our tech-

nology. It simply is not wise to cancel the building

of a great and needed dam to protect a minute fish

like the snail darter nor to damage our timber indus-

try and put thousands of workers out of jobs to pro-
red herring
tect the spotted owl. [Why fret about saving the spot-

ted owl when we have crime in our streets?] From the
sweeping generalization
early days of this great country,[all reasonable peo-
factual
ple have favored the march of progress.] Our[earliest
error
three presidents, Washington, Lincoln, and Jefferson,]

were outspoken advocates of progress.
unreliable authority
[A friend of mine who is employed by the U.S. De-

partment of Agriculture states that insecticides are

totally harmless to the public.] Yet the loud cry goes

out daily that we must ban these useful chemicals be-
arguing in a circle
cause of their threat to health. [The fact that pesti-

cides are safe argues strongly for our continued use

of them to protect the crops that feed our multitudes.]

If those who oppose pesticides could have lived in the

days when the boll weevil practically wiped out this

nation's cotton crop, they might take a different view

non sequitur

of our using chemicals. [People are living longer now than they did before we used pesticides on crops; therefore, these chemicals could not possibly be a national health hazard.] It is not the chemicals that we should worry about but those who demonstrate

post hoc, ergo propter hoc

against them. [After such a group brought pressure to ban pesticides in New Jersey, eight small farmers went into bankruptcy, and the nation was deprived of their produce.]

***either . . . or* fallacy**

Our choice is clear: [we should either make up our minds to move forward in the march of progress, ignoring all the clamor for saving or preserving this or that, or give in to these demands and give up any idea of a better way of life for ourselves and our children.] I am sure that those who read this are

flattery

[bright enough to realize that only in the first alternative is there survival and accomplishment.]

Paragraphs

41 Writing Paragraphs ⌗

Paragraphs civilize writing. Without them an essay is a wilderness of sentences in which it is easy to get lost. Writings without paragraph divisions are larger versions of fused sentences, which can be frustrating or bewildering. You must know how and where to begin and end a paragraph just as you must know how and where to begin and end a sentence. A sentence is a group of words that expresses a complete thought; **a paragraph is a group of sentences that develops a single idea.** Paragraphing and thinking, then, are inseparable.

41a Determine precisely the single idea you wish to develop in a paragraph.

To compose an effective paragraph, you need to sort out your thinking so that several ideas do not tumble about like garments in a clothes dryer. Select one appropriate thought to develop, and save others for other paragraphs. You may find useful some of the principles and methods discussed in connection with getting started on a paper (see **39a, b, c**). For illustrations of paragraphs that develop a single idea and those that ineffectually include several, see **41c**.

41b Express the central thought succinctly in a topic sentence that is effectively placed. **ts**

What the thesis statement is to an entire paper (see pp. 229–230), the **topic sentence** is to a paragraph. The topic sentence should be a clear, crisp statement of the central thought of the paragraph, and it should be placed where it will be most effective.

Sometimes you will need several attempts to produce a good topic sentence. As you write and revise, remember that you want a precise and sharply focused sentence. Below is a topic sentence from a paragraph about flags.

> Flags are interesting.

The sentence is vague and trite; it conveys little meaning. Notice the improvement in the following topic sentence.

> Traditional concerns and ideals of nations are often represented by images on their flags.

Topic sentences, then, supply direction; other sentences in the paragraph add evidence, make refinements, and develop the main idea. The following paragraph illustrates this pattern.

topic sentence   Many artists have spoken of seeing things differently while drawing and have often mentioned that drawing puts them into an altered state of awareness. In that different subjective state, they speak of feeling transported, at one with the work, able to grasp relationships that they ordinarily *body of* cannot grasp. Awareness of the passage of time fades away, *paragraph* and words recede from consciousness. They say that they feel alert and aware yet are relaxed and free of anxiety, experiencing a pleasurable, almost mystical activation of the mind.

Adapted from BETTY EDWARDS
Drawing on the Right Side of the Brain

Topic sentences most often come at the beginning of paragraphs, as above, but they can appear anywhere. Place the topic sentence wherever it will be most effective. In the following paragraph, the topic sentence is placed after the first sentence.

> The experience of missionaries in early nineteenth century Hawaii has been described elsewhere by mission historians and figures prominently in the many general histories of the islands. The missionaries' importance in island history is undeniable. The conversion of the population to Protestant Christianity, the education

system established, the provision of a Western language, the influence on chiefly leaders which resulted in a constitution, an elected assembly, and land redistribution by mid-century were striking events and critical for the future development of the island-state.

PATRICIA GRIMSHAW
Paths of Duty: American Missionary Wives in Nineteenth-Century Hawaii

If you want to set up a point with which to contrast or argue, or if you want to create a sense of suspense before you introduce the main subject, the topic sentence can come even later. It is the third sentence in the following paragraph.

> Biologists may be accustomed to working on obscure beasts—ugly horseshoe crabs, prickly sea urchins, slimy algae. But anthropologist Grover Krantz of Washington State University has a problem of a different sort. <u>He is studying, or trying to study, a creature no scientist has ever seen: the mysterious Sasquatch, or Bigfoot.</u> [The rest of the paragraph discusses the nature of the research.]

"Tracking the Sasquatch"
Newsweek

When you have several examples that you intend to use to prove a point, it is sometimes effective to present the evidence first and to end the paragraph with your central idea. In the following paragraph, the topic sentence comes at the end:

> Farmers no longer have cows, pigs, chickens, or other animals on their farms; according to the U.S. Department of Agriculture, farmers have "grain-consuming animal units" (which, according to the Tax Reform Act of 1986, are kept in "single-purpose agricultural structures," not pig pens and chicken coops). Attentive observers of the English language also learned recently that the multibillion dollar market crash of 1987 was simply a "fourth quarter equity retreat"; that airplanes do not crash—they just have "uncontrolled contact with the ground"; that janitors are really "environmental technicians"; that it was a "diagnostic misadventure of a high magnitude" which caused the death of a patient in a Philadelphia hospital, not medical malpractice; and that President Reagan was not really unconscious

while he underwent minor surgery; he was just in a "non-decision-making form." In other words, doublespeak continues to spread as the official language of public discourse.

WILLIAM LUTZ
"The World of Doublespeak"

By practicing the various placements of topic sentences, you can increase your skills as an alert writer who does not compose paragraphs automatically in one monotonous pattern.

■ Exercise 1

Underline the topic sentences in the following paragraphs and be prepared to discuss the effectiveness of their placement.

1. The world is shaped by gravity and the operations of nature depend on it. After gathering the materials of the earth into a ball, it holds them together. Opposing the titanic convulsions of the young planet, it formed the mountains. It propels the rivers and streams. Gravity pulls the rain from the clouds and flattens the surface of the sea. It gives direction to the trunks of trees and the stems of flowers.

HANS C. VON BAEYER
"Gravity"

2. The New York investment banking house of Morgan Stanley encourages people it is thinking of hiring to discuss the demands of the job with their spouses, girlfriends, or boy-friends—new recruits sometimes work 100 hours a week. The firm's managing directors and their wives take promising candidates and their spouses or companions out to dinner to bring home to them what they will face. The point is to get a person who will not be happy within Morgan's culture because of the way his family feels to eliminate himself from consideration for a job there.

RICHARD PASCALE
Fortune

■ Exercise 2

Compose two paragraphs, one with the topic sentence at the beginning and the other with the topic sentence in another position. Be prepared to justify the placement of the topic sentences.

41c Unify. Relate each sentence in a paragraph to the central idea.

Each sentence in an effective paragraph bears directly and obviously on the main point. Do not make your readers ponder and strain before they can see connections. Even slightly irrelevant material can throw the entire paragraph out of focus and lead to confusion about the direction of the argument.

The following paragraph compares secrecy to fire, but after a good topic sentence at the beginning, it includes three sentences (in italics) that stray off the point.

> Secrecy is as indispensable to human beings as fire, and as greatly feared. *Of course, we all know that fear can be a terrible barrier to communication and is generally destructive itself. Nothing is more important to society than communication.* Both fire and secrecy enhance and protect life, yet both can stifle, lay waste, spread out of control. *Naturally, gossip is also to be feared as one of the negative aspects of civilization.* Fire and secrecy can be good—as in guarding intimacy and nurturing—or bad—as in invading or consuming.

What starts out as a thoughtful and original comparison between secrecy and fire soon sprawls into a commentary on fear, communication, and gossip. Below is the paragraph as it was actually written. Notice its effective unity.

Secrecy is as indispensable to human beings as fire, and as greatly feared. Both enhance and protect life, yet both can stifle, lay waste, spread out of all control. Both may be used to guard intimacy or to invade it, to nurture or to consume. And each can be turned against itself; barriers of secrecy are set up to guard against secret plots and surreptitious prying, just as fire is used to fight fire.

SISSELA BOK
Secrets

The paragraph below attempts to develop the central idea of preserving our forests, but about halfway through, the writer abruptly changes to the topic of beauty and thus destroys the coherence of the paragraph.

An encouraging sign in modern management of national re-sources is the planting of trees systematically to replace those harvested. Large lumber companies have learned that it is in their best interests to look to the future and not merely to get what they can at the moment from the land. *Besides that, trees are beautiful and add much to the pleasure of being in nature. Only God, as the poet so aptly put it, can make a tree.* With modern tools and methods, whole forests can be destroyed in a fraction of the time that lumberjacks with their axes and handsaws attacked the woods. It is more necessary than ever, therefore, that conservation be a primary concern not only of the general citizenry but also of industry.

If the two digressive sentences printed in italics were deleted, the paragraph would be coherent and would communicate, as it should, a single well-argued point.

Often it is not so easy to make a good paragraph from a flawed one. Without planning and care, a paragraph can be merely a random collection of thoughts with only a vague central idea.

The writer of the following paragraph seems not to have thought through what he or she wanted to say.

> It is not an easy matter to be an only child. Children have differing characteristics, and some people say that infants have traits that will remain throughout their lives. As people grow up, they find that their views change, especially when they reach college. In order to understand children, we need to treat them as individuals, not as playthings. There is much truth in the saying that the child is the parent of the adult. Some writers believe that children are more mature in important ways than adults. At any rate, children do feel things keenly.

The sentences that make up the previous paragraph are not closely related and do not work toward the development of a thesis. Though the paragraph is on the general subject of children, it is so fragmented that it conveys disorder and confusion.

■ Exercise 3

Study the paragraph below, and delete all extraneous material.

As the basic social unit, the family is as important today in America as it ever was, though perhaps in a different way. Family coherence was essential in the early days of the country to ensure the survival of the individual members. They helped each other and protected each other. Today people need their families not so much to ensure physical survival as to help them through the perils of modern times, especially through such psychological perils as loss of identity. ~~America is not all bad, however. It offers the greatest freedom of all countries for individual development. America is still the land of opportunity.~~ The family gives one a sense of belonging, a sense of the past. When all else seems severed, the family can be the anchor to sanity.

41d Flesh out underdeveloped paragraphs.

Making a new paragraph after every two or three sentences, without regard to development and coherence, is nearly as distracting as having no paragraph divisions at all. Inexperienced writers often believe erroneously that the sole function of an indentation is to break the monotonous flow of print. In most expository writing, underdeveloped paragraphs are like under-nourished people: they are weak, unable to carry out their normal tasks. Make sure your paragraph divisions come only with new units of thought and that you have appropriately fleshed out an idea before going on to another one.

The following paragraphs on George Washington are ineffectively developed:

> In many ways, the version of George Washington we grew up on is accurate. He was an honest and personally unambitious politician.
>
> But history teachers rarely tell us about the complex character of the man.
>
> Most historians feel that he was not a great soldier; yet he was the luckiest man alive.
>
> He even survived the attacks of his mother. When he was president, she went so far as to state publicly that he was starving and neglecting her, neither of which was true.

Though interesting, these paragraphs are too underdeveloped to do their jobs. They are just strong enough to state the points. Contrast them with the paragraphs below:

In many ways, the version of George Washington we grew up on is accurate. He was an honest and personally unambitious politician, a devout patriot and a fearless soldier. Through circumstance, he becomes the lodestar for the swift-sailing Revolution. Even his enemies conceded that national success would have been impossible without him.

But history teachers rarely tell us about the complex character of the man. Often moody and bleak, he did not like to be touched. When he became president, he required so much ritual and formality that it caused one observer to quip: "I fear we may have exchanged George the Third for George the First." Though he drank hard with the enlisted men, he was a tough disciplinarian, describing himself as a man "who always walked on a straight line."

Most historians feel that he was not a great soldier; yet he was the luckiest man alive. He had two horses shot out from under him; felt bullets rip through his clothes and hat; and survived attacks by Indians, French and English troops, hard winters, cunning political opponents, and smallpox.

He even survived the attacks of his mother. When he was president she went so far as to state publicly that he was starving and neglecting her, neither of which was true. Then, to twist the emotional knife, she tried to persuade the new government to pass a law that future presidents not be allowed to neglect their mothers.

Adapted from DIANE ACKERMAN
"The Real George Washington"

In a few kinds of writing, short paragraphs are acceptable and expected. Newspapers, for example, generally use brief paragraphs because few details and little exposition are needed. Sometimes it is effective to use a short paragraph amid longer ones for emphasis. The length of a writing assignment to some extent may influence the length of paragraphs. A paper of one thousand words provides more room to develop full paragraphs than a short assignment does.

■ Exercise 4

Seven broad subjects for paragraphs are listed below. Choose the two that you like best, and select one aspect of each topic. On each subject, first write a skimpy paragraph. Then write a paragraph of 125 to 175 words with fuller development of the subject.

1. tabloids and sensationalism
2. preventive medicine
3. nuclear energy
4. television soap operas
5. computer technology
6. a favorite recreation
7. changing automobile designs

41e Trim and tighten sprawling paragraphs.

A sprawling paragraph is the opposite of an underdeveloped one: it needs trimming. As you write, keep alert to the length of your paragraphs. To reduce excessive length, you may need to reduce the scope of the topic sentence, but sometimes you can produce a more coherent, more sharply focused paragraph simply by discarding material. For example, all your details may be pertinent and interesting, but you may not need ten examples to illustrate your point; four or five may do it more efficiently.

41f Put the parts of a paragraph together in appropriate, coherent order. *coh*

Sentences can be arranged systematically in a number of ways. Different orders of sentences produce different effects, and what may be an appropriate arrangement for one kind of paper will not be for another. Paragraphs with no system of order are like motors with the parts haphazardly assembled: they function poorly or not at all. Study below the various patterns most often used to arrange sentences in a paragraph.

Time

In paragraphs organized by time, things that happen first usually come first in the paragraph. In many ways, this is the simplest system because chronology is a ready-made pattern. Paragraphs describing a process (how steel is made, how photographs are developed, and so forth) naturally are arranged sequentially. When steps are out of order, the paragraph is confusing. Narratives usually begin at the first event and end with the last. In some instances, however, describing a later event first and then coming back to the beginning may create a desired special effect.

The following paragraph describes events in strict chronological order:

> At the beginning of the cruise, the ship's engines seemed noisy, but the captain considered this a minor problem. He consulted with the chief engineer, who politely but firmly told him that he was imagining things. On the third day out, smoke was reported in the engine room, and the captain prepared the crew before giving orders to abandon ship. Shortly before the announcement was to be made, however, a large freighter came into view. It was an Italian ship, and few aboard spoke English, but they quickly discerned the problem, gave full assistance, and took the passengers on board their ship. Soon the fire was out and the *Ocean Wind* was in tow. After the two days that it took to reach Miami, the passengers agreed that they had never had better food, better quarters, or more attentive hosts.

Space

Descriptive paragraphs are most frequently organized according to spatial progression. If you are describing a scene or a person, you move from detail to detail of what you have observed. As with time sequence, you must maintain the logical progression, in this instance the movement of the eye as it follows the object or scene. Getting sentences out of order may resemble counting

1, 2, 5, 4, 3 instead of 1, 2, 3, 4, 5. Notice how the paragraph below carries the eye from one wall of a room to another.

> The right wall of the spacious office was covered completely with a mural of the most striking aspect. It was meant to represent a city skyline, but the buildings resembled giant trees in winter, and the color red prevailed everywhere. The rear wall was composed of one piece of glass from floor to ceiling overlooking the real skyline. The contrast was immediate, but one could also see that in a strange sense the buildings did, indeed, look like a surrealistic forest. The left wall was taken up with bookshelves filled with volumes of various shapes and sizes. A closer look revealed that the top four shelves contained all first editions of famous works. The lower four shelves, however, were given to books of only two kinds—writings on architecture and on the cultures of certain Pacific islanders.

NOTE: Paragraphs with sentences arranged by time or space sometimes do not have conventional topic sentences.

Climactic order (increasing importance)

A paragraph may also be organized by arranging sentences in order of increasing importance, or **climactic order.** If you place a lesser sentence (an anticlimax) at the end of a paragraph, you may create confusion or unintentional humor. The strategy in arranging climactically is not to put down your thoughts simply as they occur to you, but to list them first, order them according to their relative importance, and then form the paragraph. Notice how the final sentence in the following paragraph dramatically acts as a climax.

> For the poet Walt Whitman, grass became emblematic of much that fascinated him. He considered it most simply as an often overlooked and inadequately appreciated aspect of our natural surroundings. In its plentifulness it also represented to him the common people of the world and thus the ideal of democracy. It had spiritual significance for him as well, for he thought of the grass as "the uncut hair of graves," that is, as a sign of life emerging from death.

The points are ordered according to the increasing significance of grass to Walt Whitman: what it represented to him physically, intellectually, and spiritually.

General to particular, particular to general

Paragraphs in which the sentences progress from a general statement to particular details are related to deductive reasoning; those that progress from several details to a generalization are related to inductive reasoning (see **40**). Writers often compose a topic sentence and then support and explain it with details, reasons, and illustrations. This sequence moves from the general to the particular. The reverse order usually has the topic sentence at the end of the paragraph with particulars coming first. A more frequent pattern moves from the general to the particular and back to the general again.

FROM GENERAL TO PARTICULAR

> Never was there a more outrageous or more unscrupulous or more ill-informed advertising campaign than that by which the promoters for the American colonies brought settlers here. Brochures published in England in the seventeenth century, some even earlier, were full of hopeful overstatements, half-truths, and downright lies, along with some facts which nowadays surely would be the basis for a restraining order from the Federal Trade Commission. Gold and silver, fountains of youth, plenty of fish, venison without limit, all these were promised.

> DANIEL BOORSTIN
> *Democracy and Its Discontents*

FROM PARTICULAR TO GENERAL

> Photocopying makes it possible for a researcher to reproduce a long passage instantly instead of laboriously copying it in longhand or dragging along the entire book so that the section can be copied later. Ball-point and felt-point pens, now used widely instead of old-fashioned ink pens, are economical, convenient, and generally neater. The wide variety of colors in these pens enables writers to distinguish notes on one subject from those on another. Word-processing machines allow the change of single words or the revision of a line or a passage long after the original typing. Transparent tapes are now

available that can be written or typed on. A great number of conveniences have been developed for researchers and writers in the last few decades.

FROM GENERAL TO PARTICULAR TO GENERAL

Mankind's most enduring achievement is art. At its best, it reveals the nobility that coexists in human nature along with flaws and evils, and the beauty and truth it can perceive. Whether in music or architecture, literature, painting or sculpture, art opens our eyes and ears and feelings to something beyond ourselves, something we cannot experience without the artist's vision and the genius of his craft. The placing of Greek temples like the Temple of Poseidon on the promontory at Sunion outlined against the piercing blue of the Aegean Sea, Poseidon's home; the majesty of Michelangelo's sculptured figures in stone; Shakespeare's command of language and knowledge of the human soul; the intricate order of Bach, the enchantment of Mozart; the purity of Chinese monochrome pottery with the lovely names—celadon, oxblood, peach blossom, clair de lune; the exuberance of Tiepolo's ceiling where, without the picture frames to limit movement, a whole world in exquisitely beautiful colors lives and moves in the sky; the prose and poetry of all the writers from Homer to Cervantes to Jane Austen and John Keats to Dostoevsky and Chekhov—who made all these things? We—our species—did.

BARBARA TUCHMAN
Thomas Jefferson lecture, Washington, D.C.

41g   Develop a paragraph by a method that will appropriately enable it to fulfill its function.

Each paragraph has its own distinctive identity, and yet it must carry out its duty in the paper as a whole. Long-established practices may at times be helpful in planning and arranging your thoughts.

Opening and closing paragraphs   *intro/conc*

Opening and closing paragraphs of a paper are often difficult to write. The first paragraph is sometimes harder to compose than the next several pages. It must attract interest, state the purpose or thesis or argument, and then sometimes suggest the method of development that will be used in the entire paper. If the first paragraph is dull, mechanical, or obscure, the reader may decide immediately that the paper is not worth reading.

The concluding paragraph does not simply restate the opening paragraph. It does not repeat information previously discussed. Instead, it gives a brief overview of what the paper has shown and makes a final assessment of the importance and the originality of the paper in regard to its subject matter.

Methods of development

Methods of paragraph development vary with content. Several particular kinds of paragraphs are discussed below.

Definition

Whether you are using a new word or a new meaning for an old one, a definition explains a concept. It avoids the problems that arise when two persons use the same term for different things. Be as specific as possible; exemplify. Avoid definitions that are uniquely your own (unless you are willing to be challenged). Definitions, of course, may be much more elaborate than those given in dictionaries. Avoid the expression "According to Webster . . ."

> Romantic love may be described as an emotion; it is translated into behavior. A person in love wants to do something to or with his loved one. The behavior a couple settles upon comprises their love. The better the translation of emotion into behavior, the less residual emotion will remain. Paradoxically, then, people who love each other do not feel love for each other.
>
> Adapted from GEORGE W. KELLING
> *Blind Mazes: A Study of Love*

Comparison and contrast

The degree of comparison or of contrast can vary a great deal from one instance to another. You might write, for example, that two things have a great many likenesses but only one or two strong contrasts. Or the reverse could be true, with a great many contrasts but one or a few striking likenesses. Whatever your methods of recounting comparisons and contrasts, be certain that they are represented appropriately. The following paragraph on Thomas Jefferson and Abraham Lincoln begins by briefly contrasting the two men and then moves on to point out three similarities in their early lives.

> As one would expect, the formative years of Jefferson and Lincoln represent a study in contrasts, for the two men began life at opposite ends of the social and economic spectrum. There are, however, some intriguing parallels. Both men suffered the devastating loss of a parent at an early age. Jefferson's father, an able and active man to whom his son was deeply devoted, died when his son was fourteen, and Thomas was left to the care of his mother. His adolescent misogyny and his subsequent glacial silence on the subject of his mother strongly suggest that their relationship was strained. Conversely, Lincoln suffered the loss of his mother at the age of nine, and while he adored his father's second wife, he seems to have grown increasingly unable to regard his father with affection or perhaps even respect. Both Jefferson and Lincoln had the painful misfortune to experience in their youth the death of a favorite sister. And both were marked for distinction early by being elected to their respective legislatures at the age of twenty-five.
>
> DOUGLAS L. WILSON
> "What Jefferson and Lincoln Read"

Sometimes entire paragraphs may consist of either comparisons or contrasts rather than both. The following two paragraphs from the same essay simply contrast Jefferson and Lincoln.

But the differences are great. Jefferson was born into the Virginia gentry. Along with a privileged position in society, he inherited a small fortune in land and slaves. The poverty and obscurity into which Lincoln was born, on the other hand, were nearly complete. His father owned land but had great difficulty holding on to it and finally retreated with his family to southwestern Indiana, which in 1816 was little more than a wilderness, and where Abraham grew up having only the homemade clothes on his back.

In the matter of education the contrasts are equally great. Jefferson received a superb education, even by the standards of his class. It included formal schooling from the age of five, expert instruction in classical languages, two years of college, and a legal apprenticeship. Along the way he had the benefit of conspicuously learned men as his teachers—the Reverend James Maury, Dr. William Small, and George Wythe—in addition to a seat at the table of the cultivated governor of Virginia, Francis Fauquier. Lincoln had almost no formal education. Growing up with nearly illiterate parents and in an atmosphere that had, as he wrote, "nothing to excite ambition for education," Lincoln was essentially self-taught. The backwoods schools he attended very sporadically were conducted by teachers with meager qualifications. "If a straggler supposed to understand Latin, happened to sojourn in the neighborhood," Lincoln wrote, "he was looked upon as a wizzard." Jefferson read Latin from an early age and, after mastering classical languages and French, was able to teach himself Italian; Lincoln at about the same age was teaching himself grammar in order to be able to speak and write standard English.

Cause and effect

Generally a paragraph of this kind states a condition or effect and then proceeds by listing and explaining the causes. However, the first sentences may list a cause or causes and then conclude with the consequence, the effect. In either method of development, the paragraph usually begins with a phenomenon that is generally known and then moves on to the unknowns.

The following paragraph begins with an effect and proceeds to examine the causes:

This close-knit fabric [of the city] was blown apart by the automobile, and by the postwar middle-class exodus to suburbia which the mass-ownership of automobiles made possible. The automobile itself was not to blame for this development, nor was the desire for suburban living, which is obviously a genuine aspiration of many Americans. The fault lay in our failure, right up to the present time, to fashion new policies to minimize the disruptive effects of the automobile revolution. We have failed not only to tame the automobile itself, but to overhaul a property-tax system that tends to foster automotive-age sprawl, and to institute coordinated planning in the politically fragmented suburbs that have caught the brunt of the postwar building boom.

> EDMUND K. FALTERMAYER
> *Redoing America*

Example

Some topic sentences state generalizations that gain force with evidence and illustration. Proof can be provided by an extended example or several short examples, which must be acceptable as true in themselves and as representative of the generalization. Examples can add concrete interest as well as proof, as can be seen in the following paragraph:

Ours was once a forested planet. The rocky hillsides of Greece were covered with trees. Syria was known for its forests, not its deserts. Lebanon had vast cedar forests, from which the navies of Phoenicia, Persia, and Macedonia took their ship timber, and which provided the wood that Solomon used to build the temple at Jerusalem. Oak and beech forests dominated the landscapes of England and Ireland. In Germany and Sweden, bears and wolves roamed through wild forests where manicured tree farms now stand. Columbus saw the moonscape that we call Haiti "filled with trees of a thousand kinds." Exploring the east coast of North America in 1524, Verrazano wrote of "a land full of the largest forests . . . with as much beauty and delectable appearance as it would be possible to express."

> CATHERINE CAUFIELD
> "The Ancient Forest"

Classification

Paragraphs that classify explain by arranging components into groups and categories. Seeing the distinctions should lead to clear understanding of the component parts and then to understanding of the larger group or concept. **Analysis** explains one thing by naming its parts. **Synthesis** lists several categories and then puts them into a single concept or classification.

> Modern pessimism and modern fragmentation have spread in three different ways to people of our own culture and to people across the world. *Geographically,* it spread from the European mainland to England, after a time jumping the Atlantic to the United States. *Culturally,* it spread in the various disciplines from philosophy to art, to music, to general culture (the novel, poetry, drama, films), and to theology. *Socially,* it spread from the intellectuals to the educated and then through the mass media to everyone.
>
> FRANCIS A. SCHAEFFER
> *How Should We Then Live?*

Analogy

An analogy is a figurative comparison; it explains one thing in terms of another. It is likely to be most effective when you can show a resemblance not generally recognized between two things with so many differences that they are not ordinarily likened. In the paragraph below, for example, the writer convincingly reveals an analogy between human beings and lobsters:

> We are not unlike a particularly hardy crustacean. The lobster grows by developing and shedding a series of hard, protective shells. Each time it expands from within, the confining shell must be sloughed off. It is left exposed and vulnerable until, in time, a new covering grows to replace the old. With each passage from one stage of human growth to the next we, too, must shed a protective structure.
>
> Adapted from GAIL SHEEHY
> *Passages*

Process

Several methods can be used in describing a process. Most processes are given in chronological order, step by step. The kind of writing should be adapted to the particular kind of process. The simplest perhaps is the type used in a recipe, usually written in the second person or imperative mood. The necessity here is to get the steps in order and to state each step very clearly. This process tells *how to do* something. It is more instruction than exposition.

Another kind of process is the exposition of *how something works* (a clock, the human nervous system). Here the problems lie in avoiding technical terms and intricate or incomprehensible steps in the process. This kind of process is explanation; the reader may at some time need to understand it or to use it.

Still another kind of process, usually written in the past tense, tells *how something happened* (how oil was formed in the earth, how a celebration or riot began). Usually a paragraph of this type is written in the third person. It is designed to reveal how something developed (once in all time or on separate occasions). Its purpose is to explain and instruct. The following paragraph explains one theory about the process that formed the moon. The distinct steps are necessary for explanation here just as they are for instruction in a paragraph that tells how to do something.

> There were tides in the new earth long before there was an ocean. In response to the pull of the sun the molten liquids of the earth's whole surface rose in tides that rolled unhindered around the globe and only gradually slackened and diminished as the earthly shell cooled, congealed, and hardened. Those who believe that the moon is a child of earth say that during an early stage of the earth's development something happened that caused this rolling, viscid tide to gather speed and momentum and to rise to unimaginable heights. Apparently the force that created these greatest tides the earth has ever known was the force of resonance, for at this time the period of the solar tides had come to approach, then equal, the period of the free oscillation of the liquid earth. And so every sun tide was

given increased momentum by the push of the earth's oscillation, and each of the twice-daily tides was larger than the one before it. Physicists have calculated that, after 500 years of such monstrous, steadily increasing tides, those on the side toward the sun became too high for stability, and a great wave was torn away and hurled into space. But immediately, of course, the newly created satellite became subject to physical laws that sent it spinning in an orbit of its own about the earth. This is what we call the moon.

RACHEL CARSON
The Sea Around Us

41h Use transitional devices to show the relationships between the parts of your writing. *tr*

Transitional devices are connectors and direction givers. They connect content words to other words, sentences to sentences, paragraphs to paragraphs. Writings without transitions are like a strange land with no signs for travelers. Practiced writers assume that they should keep their readers informed about where a paragraph and a paper are going.

The beginnings of paragraphs can contribute materially to clarity, coherence, and the movement of the discussion. Some writers meticulously guide readers with a connector at the beginning of almost every paragraph. H. J. Muller, for example, begins a sequence of paragraphs about science as follows.

In this summary, science . . .

Yet science does . . .

Similarly the basic interests of science . . .

In other words, they are not . . .

This demonstration that even the scientist . . .

This idea will concern us . . .

In other words, facts and figures . . .

CONNECTIVE WORDS AND EXPRESSIONS

but	indeed	likewise
and	in fact	consequently
however	meanwhile	first
moreover	afterward	next
furthermore	then	in brief
on the other hand	so	to summarize
nevertheless	still	to conclude
for example	after all	similarly

DEMONSTRATIVE PRONOUNS

this   that   these   those

References to demonstratives must be clear (see p. 62).

OTHER PRONOUNS

many   each   some   others   such   either

Repeated key words, phrases, and synonyms

Repetitions and synonyms guide the reader from sentence to sentence and paragraph to paragraph.

Parallel structures

Repeating similar structural forms of a sentence can show how certain ideas within a paragraph are alike in content as well as structure. A sequence of sentences beginning with a noun subject or with the same kind of pronoun subject, a series of clauses beginning with *that* or *which,* a series of clauses beginning with a similar kind of subordinate conjunction (like *because*)—devices like these can achieve transition and show connection.

Excessive use of parallelism, however, is likely to be too oratorical, too dramatic. Used with restraint, parallel structures are excellent transitional devices.

The following paragraph on the subject of emotions illustrates how various transitional devices can help create direction and coherence.

Emotions are part of our genetic heritage. Fish swim, birds fly,
┌──── repeated words ────┐
and people feel. Sometimes we are happy; sometimes we are not.
connective word: But
But sometimes in our life we are sure to feel anger and fear, sadness

and joy, greed and guilt, lust and scorn, delight and disgust. While

we are not free to choose the emotions that arise in us, we are free

to choose how and when to express them, provided we know what
demonstrative: That
they are. That is the crux of the problem. Many people have been

parallel structure
educated out of knowing what their feelings are. When they hated,

they were told it was only dislike. When they were afraid, they were

told there was nothing to be afraid of. When they felt pain, they were

advised to be brave and smile. Many of our popular songs tell us

"pretend you are happy when you are not."

HAIM G. GINOTT
Between Parent and Child:
New Solutions to Old Problems

General exercises

■ Exercise 5

Write three paragraphs on any of the following subjects. Use at least two methods of development and two kinds of order (see pp. 283–294). Name the method you use in each paragraph.

1. spring break
2. boredom
3. poverty
4. politics and sincerity

5. shopping
6. human rights
7. rudeness
8. carelessness

■ Exercise 6

Find three good paragraphs from three different kinds of writing: from a book of nonfiction, an essay, a review, or a newspaper article. Discuss how effective paragraphs differ in different kinds of writing.

■ Exercise 7

Find an ineffective paragraph in a book or article. Analyze it in a paragraph.

Literature

42 ■ Writing About Literature, 300

42 Writing About Literature

In writing about literature, you should express your reactions in a way that does justice to the richness and the complexity of the work. A good paper is not purely subjective, although it does state your opinions. It convinces others by solid evidence derived from the work.

Although works of literature can have more than one correct interpretation, some readings can be supported more convincingly than others, and some criticisms can be shown to be erroneous.

42a Choose a literary work that interests you. Write about the feature of the work that interests you most.

Your instructor may assign you a particular work for a paper, or you may be allowed to choose for yourself. If you have any freedom of choice, select carefully. A bad selection will make success difficult even if you write well.

Give thought to your choice. You are likely to write a better paper if you choose a **work new to you**, one that no one has ever taught to you and one that you have never written about before. As much as you can, search for the kind of topic you like. Look at familiar authors, famous authors whose works you have not read, works recommended to you by people who have tastes somewhat like yours. Look at shelves of books, lists of authors and works, stories and poems you have heard about but not read, new stories and poems by authors familiar to you, anthologies, collections. You are unlikely to write a good paper about a work that you regard as dull or mediocre. If a literary work contains no mystery for you at first, you are unlikely to explain it originally or interestingly to others.

Avoid shallow and obvious literature. If you read something that makes you think, "That is exactly what I have always known," you should doubt the originality of the author. If you think, "True, but I never thought of anything like that before," you may have a good selection on your hands. A work that interprets or explains itself will not leave much for you to write about. Good papers are often derived from literature that is somewhat puzzling and obscure during and immediately after the first reading. Additional readings and careful thought can produce surprises that give you something to write about.

Do not decide to write on a particular subject before you read the work. What you think will be a good subject may not be there at all. Find your subjects as you read the literature and as you think about it afterward. Let good subjects grow and develop; bad ones will disappear or fade away. The most surprising and significant topics may come to you when you least expect them— as you slowly wake in the morning, for example, or as you engage in some kind of physical activity.

The basic methods for planning and writing a paper about literature are the same as those you would use for writing any other kind of paper (see pp. 226–229). List topics and subtopics; group them; arrange them; rearrange them; write whatever comes into your mind about them, whether it seems good or bad at the moment; write to keep the words going, even when what you are writing seems unimportant at the moment. Cross out. Put check-marks by important points; photocopy parts of works; make notations of lines, passages, images, sentences. Write quickly for a time. Think and write slowly and carefully about crucial subjects. Stop writing and walk about. Use every physical and mental device you can think of to tease good thoughts out of your mind.

Be willing to give up ideas that are inferior. Expect surprises and twists and turns during the process of writing. Few papers are written like bullets speeding toward a single idea with a calculated aim; most of the time they act more like floods, gathering valuables and debris at the same time. After the first draft, you will be able to begin separating bad from good ideas.

Kinds of literary papers

Writings about literature fall into several categories. A few of the more significant ones are explained below.

Interpretation

Many papers are interpretations, which derive mainly from close study of the literary work. An interpretative paper identifies methods and ideas. Through analysis, the writer presents specific evidence to support the interpretation.

Distinguish carefully between the thinking of a character and that of the author. Unless an author speaks in his or her own voice, you can deduce what the author thinks only from the work as a whole, from effects of events in the plot, and from attitudes that develop from good or bad characters. Some works of literature depict a character whose whole way of life is opposite to the author's views. To confuse the character with the author in this kind of work is to make a crucial mistake. Sometimes it is clear immediately what an author thinks about the characters, but not always.

Review

A good review of a book or an article

1. identifies the author, the title, and the subject.
2. summarizes accurately the information presented and the author's argument.
3. describes and perhaps categorizes the author's methods.
4. provides support (with evidence and argument) for the good points and explains the author's mistakes, errors, misjudgments, or misinterpretations.
5. evaluates the work's accomplishments and failures.

Character analysis

A character sketch is a tempting kind of paper to write, but a good analysis of a character or an explanation of an author's methods of characterization is truly a difficult task. This kind of paper easily turns into a superficial summary when it fails to consider motivations, development, change, and interrelationships of characters. Meaningful interpretation of a character in a literary work may define something that the character does not know about himself or herself, traits that the author reveals only by hints and implications. For example, if you can show that a character clearly lacks self-understanding, you may be on the way to writing a good paper.

Of course, a character analysis may *not* deal at all with *what the character is* but consider the *methods* the author uses to reveal the character. Body language and modes of speech, facial expressions or changes in mood, excessive talking or silence, and appearance are some of these methods. Remember that the author characterizes; the critic discusses the methods of characterization.

Analysis of setting

Often the time and place in which a work is set reveal important moods and meanings. Setting may help to indicate the manners and emotions of characters by showing how they interact with their environments. When you write about setting, you accomplish very little by merely describing it. Show what its functions are in the literary work.

Technical analysis

The analysis of technical elements in literature—imagery, symbolism, point of view, structure, prosody, and so on—requires special study of the technical term or concept as well as of the literary work itself. Begin by looking up the term in a good basic reference book, such as *A Handbook to Literature*. However, you should never be concerned with vocabulary and technical terms for their own sake. If you discover an aspect of a work that you wish to discuss, find the exact term for it by discussing it with your teacher, looking it up in a dictionary, and studying it in a handbook or an encyclopedia. Then determine what the technique does in the work you are writing about.

Combined approaches

Many papers combine different approaches. A thoughtful paper on imagery, for example, does more than merely point out the images, or even the kinds of images, in the literary work; it uses the imagery to interpret, analyze, or clarify something else as well—theme, structure, characterization, mood, relationship, recurrent patterns, and so on. Depending on the subject and the work you are writing about, many aspects can mingle to accomplish a single objective.

42b Give the paper a precise title.

Do not search for a fancy title at the expense of meaning. Authors of literary works often use figurative titles like *Death in the Afternoon* and *The Grapes of Wrath,* but you would be wise to designate your subject more literally. Be precise; state your subject in the title.

Do not use the title of the work as your own title, as in "Robert Frost's 'Directive.'" Do not merely announce that you are writing about a work by calling your paper "An Analysis of Frost's 'Directive,'" or "An Interpretation of . . ." or "A Criticism of . . ." The fact that you are writing the paper indicates that you are writing an analysis, a criticism, or something of the sort.

Stick to the topic named in the title. The topic sentence of every paragraph should point back to the title, the introduction, and the thesis statement.

42c Organize and develop the paper according to significant ideas.

Do not automatically organize your paper by following the sequence of the literary work. Sometimes the result of this order can be poor topic sentences, summary rather than analysis, mechanical organization, and repetitive transitional phrases.

Usually it is better to break up your overall argument into several aspects and to move from one of these to the next.

First sentences that provide mechanical and dull information do not encourage further reading. Avoid generality, as in the following sentence:

> T. S. Eliot, in his famous poem written in 1922, *The Waste Land,* expressed a theme that has been a frequent subject of works of literature.

Instead, you might try something like this:

> The relationship between the physical, mental, and spiritual health of a ruler of a nation and the condition of the people is a subject of T. S. Eliot's poem *The Waste Land.*

The first sentence states nothing of true significance; the second announces a particular topic to be explored.

Just as a paper should not begin mechanically, the parts should not contain mechanical first sentences and dull transitions. Your paragraphs will not attract readers if they use beginnings like these:

"In the first stanza . . ."
"In the following part . . ."
"At the conclusion . . ."

42d Do not summarize and paraphrase excessively.

A certain amount of summarizing is usually necessary. To a slight extent, summary can involve interpretation. But tell the story, quote, and paraphrase only enough to prove your point. Mere detailed summary is inadequate; summary must prove your argument, not be an end in itself. When you paraphrase, distinguish clearly between the author's thinking and your ideas about what he or she has written.

NOTE: When you do summarize, use the historical present tense (see **4c**). Tell what *happens,* not what *happened.*

42e Think for yourself.

The purpose of your paper is to *state your opinions* and to *provide evidence.* If you find that you have no worthwhile opinions of your own about the meaning or the art of the work, find a literary work that stimulates your thought.

42f Write about the literature, not about yourself or "the reader."

The process by which you discovered what you are writing about is usually a dull subject to other people. Give the results of your explorations, not details of the various steps.

Your teacher will know that what you write is your opinion unless you indicate otherwise; therefore, you should not repeat expressions like "In my opinion" or "I believe that." Extensive use of first-person pronouns (*I, my*) takes the emphasis from what you are writing about.

In an effort to avoid the pronoun *I,* do not develop the annoying habit of repeating the phrase "the reader." Writing about "the reader" is logically erroneous because the expression implies readers in general, and you cannot speak for this vast, varied, and complex group. Once you begin using "the reader," it is difficult to stop, and your style will be greatly weakened by the resulting wordiness. Though you may not realize it, when you make extensive use of first-person pronouns and "the reader," you are creating an impression of self-consciousness that can annoy or even alienate.

The rough draft entitled "An Analysis of 'What You See Is the Real You'" (pp. 239–243) vividly illustrates the problems discussed above. Review that paper, noting how the student's use of "I" and "the reader" severely damages style. Then note how the faults were corrected in the model paper on pp. 255–257.

42g Provide enough evidence to support your ideas.

Strike a proper balance between generalizations and detailed support of your points. Make a point, develop its specifics and ramifications, quote the work, and show how the point is supported by the quotation. Avoid long quotations.

Papers, and even paragraphs, usually should not begin with a quotation. Readers prefer to see what you have to say first.

42h Do not moralize.

Good criticism does not preach. Do not use your paper as a platform from which to state your views on the rights and wrongs and conditions of the world. A literary paper can be spoiled by an attempt to teach a moral lesson.

Begin your paper by introducing the work and your ideas about it as literature. If the author discusses morals, religion, or social causes, his or her literary treatment of these subjects is a proper topic. Do not begin your paper with your own ideas about the world and then attempt to fit the literature to them or merge your ideas with the author's. Think for yourself, but *think about the author's writing.*

42i   Acknowledge your sources.

Define the difference between what other critics have written and what you think. State your contribution. Do not begin papers or paragraphs with the names of critics and their views before you have presented your own ideas.

Develop your own thesis. Stress your views, not those of others. Use sources to show that other critics have interpreted the literature correctly, to correct errors in criticism that is otherwise excellent, to show that a critic is right but that something needs to be added, or to show that no one has previously written on your subject.

It is not a problem if no critic has written about the work or the point you are making unless your instructor requires you to have a number of sources. However, it is a serious error to state incorrectly that nothing has been written about your subject. Be thorough in your investigation.

For bibliographies of writings about literature, see **43b**. For information about plagiarism and documentation, see **43f** and **43g**.

Writing a paper about a poem

Suppose you are writing about Mary Oliver's "The Black Walnut Tree." The text of the poem is printed below. Read it through. Read it again and think of or list topics you might want to discuss. Look at the questions that follow the poem. Study the poem as you think of the general questions and your own list of topics. At some stage in this process, write down answers, take notes, and prepare to write a critical paper.

The Black Walnut Tree

My mother and I debate:
we could sell
the black walnut tree
to the lumberman,
and pay off the mortgage.
Likely some storm anyway
will churn down its dark boughs,
smashing the house. We talk
slowly, two women trying
in a difficult time to be wise.
Roots in the cellar drains,
I say, and she replies
that the leaves are getting heavier
every year, and the fruit
harder to gather away.
But something brighter than money
moves in our blood—an edge
sharp and quick as a trowel
that wants us to dig and sow.
So we talk, but we don't do
anything. That night I dream
of my fathers out of Bohemia
filling the blue fields
of fresh and generous Ohio
with leaves and vines and orchards.
What my mother and I both know
is that we'd crawl with shame
in the emptiness we'd made
in our own and our fathers' backyard.
So the black walnut tree
swings through another year
of sun and leaping winds,
of leaves and bounding fruit,
and, month after month, the whip-
crack of the mortgage.

MARY OLIVER

Questions and topics for consideration

1. What is the significance of the two characters?
2. What are the roles of a mother and a daughter? Would two sisters have the same meaning and poetic effect as characters?
3. The poem does not name the subjects of the women's talk. What are some of the possibilities?
4. What is the importance of the characters' fear that the tree may be destroyed by a storm?
5. Explain the women's concerns about the roots, the cellar drains, the leaves, the fruit (or nuts).
6. Interpret the "sharp and quick edge."
7. What is the meaning of digging and sowing?
8. Explain the daughter's dream about Bohemian fathers.
9. What does "emptiness" mean?
10. Explain the ending of the poem—the winds, leaves, and fruit and also the mortgage.
11. Consider the meaning of the tree, the character of the people, and the nature of their problems.
12. Why did the poet choose a black walnut tree? Why does the poem use the word *fruit* for nuts? Why would a lumberman be interested in a black walnut? (You may know the answer, or you may need to do research.)

■ Exercise 1

Using your answers to the questions above, write a paper about "The Black Walnut Tree."

Writing a paper about a work of fiction

Read the following short story carefully.

On the U.S.S. *Fortitude*

Ron Carlson

Some nights it gets lonely here on the U.S.S. *Fortitude*. I wipe everything down and sweep the passageways, I polish all the brass and check the turbines, and I stand up here on the bridge charting the course and watching the stars appear. This is a big ship for a single-parent family, and it's certainly better than our one small room in the Hotel Atlantis, on West Twenty-second Street. There the door wouldn't close and the window wouldn't open. Here the kids have room to move around, fresh sea air, and their own F/A-18 Hornets.

I can see Dennis now on the radar screen. He's out two hundred miles and closing, and it looks like he's with a couple of friends. I'll be able to identify them in a moment. I worry when Cherry doesn't come right home when it starts to get dark. She's only twelve. She's still out tonight, and here it is almost twenty-one hundred hours. If she's gotten vertigo or had to eject into the South China Sea, I'll just be sick. Even though it's summer, that water is cold.

There's Dennis. I can see his wing lights blinking in the distance. There are two planes with him, and I'll wait for his flyby. No sign of Cherry. I check the radar: nothing. Dennis's two friends are modified MIGs, ugly little planes that roar by like the A train, but the boys in them smile and I wave thumbs up.

These kids, they don't have any respect for the equipment. They land so hard and in such a hurry—one, two, three. Before I can get below, they've climbed out of their jets, throwing their helmets on the deck, and are going down to Dennis's quarters. "Hold it right there!" I call. It's the same old story. "Pick up your gear, boys." Dennis brings his friends over—two nice Chinese boys, who smile and bow. "Now, I'm glad you're here," I tell them. "But we do things a certain way on the U.S.S. *Fortitude*. I don't know what they do where you come from, but we pick up our helmets and we don't leave our aircraft scattered like that on the end of the flight deck."

"Oh, Mom," Dennis groans.

"Don't 'Oh, Mom' me," I tell him. "Cherry isn't home yet, and she needs plenty of room to land. Before you go to your quarters, park these jets below. When Cherry gets here, we'll have some chow. I've got a roast on."

I watch them drag their feet over to their planes, hop in, and begin to move them over to the elevator. It's not as if I asked him to clean the engine room. He can take care of his own aircraft. As a mother, I've learned that doing the right thing sometimes means getting cursed by your kids. It's O.K. by me. They can love me later. Dennis is not a bad kid; he'd just rather fly than clean up.

Cherry still isn't on the screen. I'll give her fifteen minutes and then get on the horn. I can't remember who else is out here. Two weeks ago, there was a family from Newark on the U.S.S. *Tenth Amendment,* but they were headed for Perth. We talked for hours on the radio, and the skipper, a nice woman, told me how to get stubborn skid marks off the flight deck. If you're not watching, they can build up in a hurry and make a tarry mess.

I still hope to run across Beth, my neighbor from the Hotel Atlantis. She was one of the first to get a carrier, the U.S.S. *Domestic Tranquillity,* and she's somewhere in the Indian Ocean. Her four girls would just be learning to fly now. That's such a special time. We'd have so much to talk about. I could tell her to make sure the girls always aim for the third arresting wire, so they won't hit low or overshoot into the drink. I'd tell her about how mad Dennis was the first time I hoisted him back up, dripping like a puppy, after he'd come in high and skidded off the bow. Beth and I could laugh about that—about Dennis scowling at his dear mother as I picked him up. He was wet and humiliated, but he knew I'd be there. A mother's job is to be in the rescue chopper and still get the frown.

I frowned at my mother plenty. There wasn't much time for anything else. She and Dad had a little store and I ran orders and errands, and I mean ran—time was important. I remember cutting through the Park, some little bag of medicine in my hand, and watching people at play. What a thing. I'd be taking two bottles of Pepto-Bismol up to Ninety-first Street, cutting through the Park, and there would be people playing tennis. I didn't have time to stop and figure it out. My mother would be waiting back at the store with a bag of crackers and cough medicine for me to run over to Murray Hill. But I looked. Tennis. Four people in short pants standing inside that fence, playing a game. Later, I read about tennis in the paper. But tennis is a hard game to read about at first, and it seemed a code, like so many things in my life back then, and what did it matter, anyway? I was dreaming, as my mother was happy to let me know.

But I made myself a little promise then, and I thought about it as the years passed. There was something about tennis—playing inside that fence, between those lines. I think at first I liked the idea of limits. Later, when Dennis was six or so and he started going down the block by himself, I'd watch from in front of the Atlantis, a hotel without a stoop—without an entryway or a lobby, really—and I could see him weave in and out of the sidewalk traffic for a while, and then he'd be out of sight amid the parked cars and the shopping carts and the cardboard tables of jewelry for sale. Cherry would be pulling at my hand. I had to let him go, explore on his own. But the tension in my neck wouldn't release until I'd see his red suspenders coming back. His expression then would be that of a pro, a tour guide—someone who had been around this block before.

If a person could see and understand the way one thing leads to another in this life, a person could make some plans. As it was, I'd hardly even seen the stars before, and now here, in the ocean, they lie above us in sheets. I know the names of thirty constellations, and so do my children. Sometimes I think of my life in the city, and it seems like someone else's history, someone I kind of knew but didn't understand. But these are the days; a woman gets a carrier and two kids in their Hornets and the ocean night and day, and she's got her hands full. It's a life.

And now, since we've been out here, I've been playing a little tennis with the kids. Why not? We striped a beautiful court onto the deck, and we've set up stanchions and a net. I picked up some racquets three months ago in Madagascar, vintage T-2000s, which is what Jimmy Connors used. When the wind is calm we go out there and practice, and Cherry is getting quite good. I've developed a fair backhand, and I can keep the ball in play. Dennis hits it too hard, but what can you do—he's a growing boy. At some point, we'll come across Beth, on the *Tranquillity,* and maybe all of us will play tennis. With her four girls, we could have a tournament. Or maybe we'll hop over to her carrier and just visit. The kids don't know it yet, but I'm learning to fly high-performance aircraft. Sometimes when they're gone in the after-noons, I set the *Fortitude* into a wind at thirty knots and practice touch and go's. There is going to be something on Dennis's face when he sees his mother take off in a Hornet.

Cherry suddenly appears at the edge of the radar screen. A mother always wants her children somewhere on that screen. The radio crackles. "Mom. Mom. Come in, Mom." Your daughter's voice, always a sweet thing to hear. But I'm not going to pick up right away. She can't fly around all night and get her old mom just like that.

"Mom, on the *Fortitude.* Come in, Mom. This is Cherry. Over."

"Cherry, this is your mother. Over."

"Ah, don't be mad." She's out there seventy-five, a hundred miles, and she can tell I'm mad.

"Cherry, this is your mother on the *Fortitude.* You're grounded. Over."

"Ah, Mom! Come on. I can explain."

"Cherry, I know you couldn't see it getting dark from ten thousand feet, but I also know you're wearing your Swatch. You just get your tail over here right now. Don't bother flying by. Just come on in and stow your plane. The roast has been done an hour. I'm going below now to steam the broccoli."

Tomorrow, I'll have her start painting the superstructure. There's a lot of painting on a ship this size. That'll teach her to watch what time it is.

As I climb below, I catch a glimpse of her lights and stop to watch her land. It's typical Cherry. She makes a short, shallow turn, rather than circling and doing it right, and she comes in fast, slapping hard and screeching in the cable, leaving two yards of rubber on the deck. Kids.

I take a deep breath. It's dark now here on the U.S.S. *Fortitude*. The running lights glow in the sea air. The wake brims behind us. As Cherry turns to park on the elevator, I see that her starboard Sidewinder is missing. Sometimes you feel that you're wasting your breath. How many times have we gone over this? If she's old enough to fly, she's old enough to keep track of her missiles. But she's been warned, so it's O.K. by me. We've got plenty of paint. And, as I said, this is a big ship.

Some final hints on writing about literature

Read the work once carefully.

Decide on a subject generally.

Narrow the subject as much as you can before a second reading.

Ask yourself questions as you read the work a second time, a third, and so on.

Write down the answers to your questions.

Arrange the answers by topics.

Decide on a final approach, topic, title, and thesis.

Discard the erroneous answers to your questions.

Discard the irrelevant points. Choose the one you wish to discuss.

Select the best topics.

Put them in logical order.

Rearrange them as you carry out the creative process of thinking through the subject and writing the paper.

Revise as many times as necessary.

Do not display or emphasize technical terms, which may sometimes be as distracting as they are helpful. Your reader will not be impressed by attempts to show off. That is no substitute for straight thinking and clear writing.

Be sure early in the paper to name the work that you write about
and the author (even though you may have this information
in the title). Do not misspell the author's name—as often
happens. Ernest Hemingway, for example, spelled his last
name with only one *m*.

■ Exercise 2

*Write down a list of questions about "On the U.S.S. Fortitude" like
those about "The Black Walnut Tree." Make a list of topics for
papers. Write a paper about "On the U.S.S. Fortitude" before you
read the model paper that follows.*

Model paper

Dealing with the Darkness: Ron Carlson's

"On the U.S.S. Fortitude"

The narrator of Ron Carlson's short story "On the U.S.S. Fortitude" is a woman with a great responsibility. Hers is "a single-parent family," she says— no husband to help her support and guide her two children. The story is a glimpse into her mind that reveals how she sees her responsibility as a mother and how she copes with it, especially how she deals with the threats represented by darkness.

Nighttime, the setting of the story, is greatly on the narrator's mind. She states in the first sentence: "Some nights it gets lonely here on the U.S.S. Fortitude." The nights bother her most because then she feels most keenly her loneliness. The darkness brings more than loneliness, however: it carries the threat of danger to her children. They are still young and highly vulnerable. Though the older child, Dennis, has become confident that he can handle himself in any situation and the daughter, Cherry, is starting to be impatient with her mother's worrying, still they need a strong hand to direct and correct them, to bring them through the darkness, which is emphasized throughout the story.

Through imagination, fortitude, and discipline, the mother copes with the darkness that represents loneliness to her and danger to her children. The most apparent result of her vivid imagination is her fantasy of being the captain of an aircraft carrier. In a metaphor that is both impressively sustained and humorous, she thinks of their new living quarters—no longer cramped as they were at the Hotel Atlantis in the city—as the carrier U.S.S. Fortitude, her chil-dren as naval fighter pilots, and herself as the ship's commander plotting the whereabouts of the pi-lots on radar and awaiting with some anxiety and im-patience their return to the ship. In a sense she is playing a game, dreaming. When she was a child and became interested in tennis and its codelike language, her mother called her a dreamer, and so she is. But her dreaming, her active imagination, has provided her with a way of coping with the difficult circumstances in her life. The military metaphor she imaginatively creates works well in the story because every mother fights a kind of war in bringing up her children, battling the forces of darkness to keep them safe and to educate them properly.

The narrator has designated her imaginary carrier the Fortitude, an appropriate name since that is a quality she strongly exhibits. She knows that courage, determination, and endurance—the ingredients of fortitude—are all necessary not only for a military commander but also for a mother trying to raise her children alone. She has much in common with the captain of a naval aircraft carrier. Obviously fortitude is necessary to engage in the military campaigns of war, but it also takes fortitude for a parent to fight off the threats of darkness to her children. "Doing the right thing," says the narrator of "On the U.S.S. Fortitude," is difficult because it "sometimes means getting cursed by your kids."

Nevertheless, she will courageously do the right thing because of her self-discipline and because she

4

realizes the importance of discipline in the lives of her children. "They can love me later," she says, sounding like a military leader who trains complaining troops hard for combat. She will make them follow orders and will punish them when they do not. Discipline, she feels, will help them safely through the darkness. When she was a little girl, she was attracted to the game of tennis because, as she puts it, "I liked the idea of limits." Though she does not try to hold her children back and prevent them from maturing, she stresses the importance of limits, of lines drawn beyond which they should not go. Her rule of life is to play the game well within the lines and to teach her children to do the same thing. That does not mean living without adventure or delight, for she has herself started flying, her way of suggesting that she is taking on new and different ventures.

In "On the U.S.S. Fortitude" Ron Carlson has written a story of hope. Though the mother-narrator has gone through some hard times, stars are now appearing in the night. Her imagination, fortitude, and discipline have proved to be her stays against the darkness. She has developed a sustained comparison between her role and that of a military commander, but throughout the story the captain in her is never more important than the mother. If she charts the course, she also "wipe[s] everything down and sweep[s] the passageways." If she is the commander, she is also, as she makes clear to Cherry, "your mother on the Fortitude."

■ Exercise 3

 1. *Write a brief description of the methods, techniques, and subject of criticism in the model paper.*
 2. *Suggest other topics for critical papers about Carlson's story.*

■ Exercise 4

Write a critical paper about a short story assigned or approved by your instructor.

Writing a paper about a film

Most of the principles and suggestions presented in the chapter on writing about literature will help you in writing about films. You should always keep in mind, however, that film is a visual art with its own advantages and problems distinct from those of short stories, novels, poems, and even drama. In pursuing your project, be prepared to see the motion picture several times and to study parts of it in even more detail. A single viewing will seldom produce a perceptive and interesting paper.

The phases of your own composing and writing processes about films may vary considerably both in nature and in order. One possible way of proceeding is suggested in the following steps:

 1. Select a film to write about *soon* after the paper is assigned. Choose a particular film because you have an interest in the subject matter or the film's reputation.
 2. View the entire film on a video cassette recorder twice to determine its chief meanings and techniques. Jot down notes and random ideas.

3. From these notes and from brainstorming, arrive at a tentative topic, one narrow enough to cover in a few pages but important enough to relate to larger areas of interest.
4. Identify scenes in the film that pertain to the topic and review each of these several times while making notes.
5. After reviewing all your notes and thinking through the possibilities, decide whether the topic is promising.
6. Read generally about the film's subject, production, and merits. (*Note:* Research for particular assignments is often neither required nor prohibited. Follow your teacher's instructions in the use of secondary material.)
7. Organize an argument based on the contribution of your subject to such matters as the film's tone, characterization, and theme.
8. Write a rough draft.
9. Revise extensively, trying to achieve coherence as well as clarity and correctness.
10. Type the paper and proofread it carefully.

Model paper

Patton's Dog

When the film Patton was released in 1970, the
public hailed it as one of the great motion pictures
of modern times. The Academy of Motion Picture Arts
and Sciences was impressed: Patton received seven
academy awards, including those for best picture, best
director (Franklin J. Schaffner), best actor (George
C. Scott, though he refused it), and best screenplay.
The script was written by Francis Ford Coppola, who,
after doing extensive research on Patton's life, decided neither to glorify Patton nor to condemn him but
to portray his complexity with understanding. To help
do that, he introduced into the film a bull terrier,
who played Patton's dog, Willie.

Though Patton has been the subject of extensive
critical writing, the function of Patton's dog has
gone largely unnoticed. The extent of Coppola's research into Patton's life is evident here. Patton
really did have a white bull terrier named Willie that
had originally belonged to a pilot in the RAF killed
in combat. However, Coppola created eleven scenes

with the dog to bring out comedy, character, and theme.

Before the debut of Willie, Patton is the focus of several scenes of action, such as in the battle of El Guettar. Moments later, he climbs out a window at his headquarters in North Africa to stand boldly in the path of two German planes strafing the city. Scenes like these steadily emphasize conflict and violence until a change of pace is needed. It comes in the form of comic relief named Willie. After Patton arrives in a staff car to give a speech for a women's organization, his aide, Colonel Codman, opens the door and Patton appears with Willie on a leash. Describing his new dog to Codman, Patton says: "Bred for combat. I'm gonna call him William—for William the Conqueror." As Patton and Codman walk toward the podium, Patton's dog begins to bark at a small dog belonging to one of the women. Patton says, "Watch this, Codman," who replies, "General, he'll kill that dog." Smilingly, Patton says, "Naw, I'll hold him." As the woman's dog begins to bark ferociously, Willie,

3

"bred for combat," retreats in fear. The woman picks
up her lap dog and walks toward Patton with an apol-
ogy: "I'm terribly sorry, General. Did Abigail
frighten your dog?" Patton reassures her with a po-
lite but embarrassed smile, but when she leaves, he
looks down disgustedly at his dog and says: "Your name
isn't William; it's Willie." It is one of the truly
comic moments in the film.

Willie's function as comic relief continues into
the next scene, where positioning of the camera
clearly illustrates how a film can create effects
hardly possible in other art forms. Just as George C.
Scott is about to deliver the speech Patton made to a
gathering of British women, a speech that caused much
consternation and seriously damaged the general's
military career, the camera shifts abruptly to Willie
looking sheepishly at Abigail, the same small dog that
had terrorized him just a few moments before. She
reacts to Willie's stare by barking at him, and Willie
once again retreats in fear. The camera then shifts
back to Patton as he makes his notorious prediction

4

that America and Britain would be the sole rulers of the postwar world, with Russia left out of the picture. Willie's comic performance in this scene prevents it from becoming an all-too-early climax in a film about Patton's career, which still has a long way to go. In addition, Patton's speech, received with high seriousness at the time it was actually given, contains in historical perspective an element of the comic. The film underscores this aspect by associating the speech with Willie's humorous performance on the podium.

In addition to furnishing a much-needed comic ingredient, Patton's dog is invaluable as a foil to the general's character: Willie is constantly Patton's opposite. In the two scenes with Abigail as well as later in the film when he is badly startled by a cart pulled by horses, Willie manifests a trait that Patton finds intolerable: fearfulness. In fact, Patton was severely disciplined during World War II for slapping a soldier with battle fatigue, a condition that the general considered an excuse for cowardice. If Willie

5

has no stomach for fighting, he also contrasts with his owner physically. George C. Scott projects Patton as aristocratically straight in posture and well co-ordinated in movements, while Willie is awkward and lovingly clumsy. Unlike Patton, Willie seems to have no enemies (except Abigail), and he is friendly to all he encounters. He sits in the lap of a sergeant while on a plane headed for France. His manner appears to express kindness and perhaps sympathy toward the American soldiers walking by him as they go to rescue troops at Bastogne. In contrast, Patton's demeanor is prideful as he thinks of his troops—they are merely parts in his war machine, noble and fine, but never-theless expendable. Willie's mildness and good humor are evident in a scene when he approaches his master, tail wagging, in Patton's headquarters, but the gen-eral has no time for such simple affection: he is studying the art of war.

Although the dog serves generally to help bring out Patton's commitment to war as a way of life, Wil-lie ironically is also an instrument through which the warmonger is shown to have a tender but sensitive side. For example, in a scene where Patton is in his headquarters after the liberation of Paris, he walks past Willie (sitting like a human being in a chair) and pets him with what seems genuine affection. Despite Willie's failure to live up to his breeding as a fighter, Patton—this man who thrives on fighting—keeps him by his side, a startling fact that suggests effectively the general's great complexity.

Among other things, Patton is a motion picture with a powerful but subtle theme, one that has resounded through the ages: those whose values endure are the peacemakers, not the conquerors. Yet Patton is not a propagandistic or preachy film. This underlying theme is unobtrusively developed largely through a series of images and references relating to the figure of the conqueror and, more specifically, to the historical William the Conqueror. Patton originally called his dog William the Conqueror because of his ardent admiration for that great combatant. The film sustains a series of allusions to William. For example, Patton at one point is seen reading the <u>History of the Norman Conquest</u> and formulating his battle plans in Europe as he thinks of William's tactics. He imagines himself to be in the line of the world's great conquerors; he even suspects that he is related to them through reincarnation. But it is his love of war and warriors and his image of himself as a twentieth-century William the Conqueror that finally brings him down from the majestic if profane conqueror who opens the film with a speech to his troops to the dejected outcast of the final scene. There he walks behind his dog, Willie, as his voice off-screen describes nostalgically the receptions of ancient conquerors as they returned in victory to their cities. Only Patton and his dog are present in this last scene; Willie leads the way to suggest perhaps that the true conquerors in life are not the Pattons but the peace-loving meek. Ironically, Willie was the conqueror after all.

Research

43 Writing the Research Paper

Research is the act of gathering new information. In setting out on a research project, your purpose should not be merely to find enough material on a subject to satisfy your instructor that you have used the library and know how to find, quote from, and cite sources. The primary reason for research is to learn enough about a subject to write or speak about it knowledgeably. Research is necessary if you are to know what you are talking about.

After discovering new information on your subject and examining and evaluating various ideas, you will arrive at a crucial point: your basic research is finished, but the project is not. A good research paper is not simply a series of paraphrases, citations, and quotations. Formulate your own thesis, give your own slant to the subject, think through questions, and present some conclusions of your own. A research paper embodies the results of exploration into your own thinking as well as into the thinking of others.

43a Select a subject that interests you. Limit it to a manageable size.

Writing a research paper usually involves an extended period of time and a variety of activities and skills. If you get an early start (delay is deadly on a research project), give adequate consideration to potential topics, and choose one that captures your imagination and arouses your curiosity, you will find the process not only instructive but also intriguing. In choosing a subject, you may find it helpful to spend time thumbing through current periodicals. Ask your reference librarian to point out bibliographic guides such as *Hot Topics* and *Editorial Research Reports.* Make sure, however, that you carry out *controlled browsing,* for you can waste time if you do not constantly keep in mind that even at this stage you are doing research; you are searching for a good topic, not merely reading and looking randomly.

Choose a general subject area

Begin the process of selecting a topic by asking yourself what truly interests you. Perhaps you would like to learn more about photography, the fast-food industry, or the Roman Empire. The possibilities are infinite. Free your imagination and let it roam. If nothing comes to mind and browsing in magazines and a good encyclopedia produces no positive results, turn to a list of broad areas such as the following, decide on one you like, and then go on to narrow it down.

advertising	acting	law	business
religion	literature	art	oceanography
ecology	philosophy	history	government
science	anthropology	economics	medicine
geography	industry	sociology	archaeology
psychology	folklore	agriculture	education

Get more specific

Suppose you have chosen medicine for your general area. Naturally, you will need to focus on one aspect of this vast field. To help yourself decide on an aspect of the area you have selected on the list, draw a circle around it and then review the other entries on the list or add more topics that represent your secondary interests. Draw arrows from your primary area to the secondary ones, as in the example below:

advertising	acting	law	business
religion	literature	art	oceanography
ecology	philosophy	history	government
science	anthropology	economics	medicine
geography	industry	sociology	archaeology
psychology	folklore	agriculture	education

By associating a major area of interest with secondary (but vital) concerns, you move closer to a subject. For instance, think of aspects of medicine that involve law and sociology. After more thinking and general reading, a specific topic may emerge, such as "The Effect of Malpractice Suits on the Public's Image of the Physician."

With an entirely different set of interests, you might circle, say, *science* and then draw arrows to *history, geography, anthropology, archaeology,* and *ecology.* Many possibilities for a good topic exist in such a combination, but this procedure brings to light a pattern of interest that gets you closer to your final subject (which might be in this instance "What Happened to the Dinosaurs: A Continuing Debate"). The student who wrote the model paper on pp. 371–419 circled *oceanography* and then drew arrows to *psychology, science,* and *sociology.* Some hard thinking then resulted in a good subject, "The Dolphin Controversy."

Avoid inappropriate subjects

After you have tentatively selected a subject, trim it, refine it, and polish it into a definite title and then ask yourself the following questions about it.

1. *Is it too broad?* If you set out to write a research paper of two thousand words on "Religions of the World," the paper will probably be too general, undetailed, and shallow. You might be very much interested in "East Africa: 1824–1886," but this is a subject for a book. The broad scope of such titles is a clue that you will need to narrow your focus to find a workable subject.

2. *Is it too technical?* Consider this question from two angles: yours and your audience's. You may know a great deal about "Bronchial Morphometry in Odontocete Cetaceans," but an audience of general readers will not. Try to reconstitute your subject into one that will be more understandable to your readers.

3. *Does it promise more than it can deliver?* Examine your subject carefully to be sure that the promise implicit in the title can be fulfilled. For example, how could a research paper live up to the expectations created by a subject such as "Is There a God?" Even a topic that seems sufficiently limited and unspecialized can be inappropriate because it is overly ambitious.

4. *Is it too close to home?* Sometimes controversial issues create such an emotional response in the writer that a sustained, even tone is impossible. If volatile subjects such as abortion, television ministers, racism, and government-sponsored lotteries cause you such anguish that you cannot be calm and objective, avoid them. Interest is one thing; subjective indulgence is another. Controversial issues can be excellent and compelling topics for papers, but be certain that you are not so deeply involved that you will get carried away emotionally.

5. *Is it dull?* You must stand outside yourself to answer this question. You may be fascinated with "Four Types of Bathroom Tubs," but ask yourself if others will be. Usually you can relate what might be an uninteresting topic to other concerns and produce a promising subject. If for some reason you are intrigued with bathtubs, you might develop an imaginative subject by linking that interest with art ("The Bathtub in Postmodern Impressionism") or with psychology ("Psychological Implications of Modern Bathtub Designs").

43b Follow a search strategy for using the library.

Once you have determined your subject, you should compile a **working bibliography**, a list of publications pertinent to your topic. The items on this list should include only the author's name, the title, and any information you need to find the source in the library. Where to begin and how to proceed in your search are matters that require planning; otherwise, you can waste countless hours in random and fruitless excursions. Usually the best place to begin is the **reference room**. Your teachers in this area are the reference librarians, indispensable professionals in any college or university library. They can save you time by suggesting shortcuts or other alterations in your search strategy; they can also direct you to helpful guides, such as the *Library of Congress Subject Headings,* which enables you to find references under subject headings in the library catalog.

General reference aids

Before you turn to the library's catalog, however, look up your subject in a general or specialized encyclopedia, dictionary, or other such reference work. This is often an excellent starting point because the entries are usually general enough to give you a broad outline of information. For example, the article on dolphins in the *Encyclopedia Americana* (1986) gives details about the physical makeup of the species, their eating habits, and their reproductive cycle. It then concentrates on the most familiar variety, the bottlenose dolphin. It provides historical data on the relationship between human beings and dolphins, and ends with a summary of recent findings about the species' intelligence, ability to imitate human speech, and extensive use of a sonarlike faculty.

Sometimes articles in encyclopedias conclude with suggestions for further reading (a **bibliography**), which can start you on your way toward compiling a list of materials to read or consult. Below is a listing of encyclopedias, dictionaries, bibliographies, and other reference aids that you will find useful in the early stages of your research. Ask your reference librarian about additional sources.

GENERAL REFERENCE AIDS

Articles on American Literature. 1900–1950; 1950–1967; 1968–1975.

Cambridge Bibliography of English Literature. 1941–1957. 5 vols.; *New Cambridge Bibliography of English Literature,* 1969–1977. 5 vols.

Cambridge Histories. Ancient, 12 vols., rev. ed. in progress; Medieval, 8 vols.; Modern, 13 vols.; *New Cambridge Modern History.* 14 vols.

Collier's Encyclopedia.

Columbia Lippincott Gazetteer of the World. 1962.

Contemporary Authors. 1962–.

Current Biography. 1940–. "Who's News and Why."

Dictionary of American Biography. 1928–1937. 20 vols. 8 supplements, 1944–1990.

Dictionary of American History. rev. ed., 1976–1978. 8 vols.

Dictionary of National Biography. 1885–1901. 22 vols. 9 supplements, 1912–1986.

Encyclopedia Americana. Supplemented by the *Americana Annual.* 1923–.

Encyclopedia Judaica. 1972. 16 vols. Supplemented by its *Yearbook.*

Encyclopedia of Philosophy. 1967. 8 vols.

Encyclopedia of Religion. 1987. 15 vols.

Encyclopedia of Religion and Ethics. 1908–1927. 13 vols.

Encyclopedia of World Art. 1959–1968. 15 vols. 2 supplements, 1983–1987.

Encyclopedia of World History. 5th ed., 1972.

Essay and General Literature Index. 1900–. "An Index to . . . Volumes of Collections of Essays and Miscellaneous Works."

Facts on File; a Weekly World News Digest . . . 1940–.

Harvard Encyclopedia of American Ethnic Groups. 1980.

Information Please Almanac. 1947–.

International Encyclopedia of the Social Sciences. 1968. 17 vols. *Biographical Supplement.* 1979.

Literary History of the United States. 1974. 2 vols.

MLA International Bibliography of Books and Articles on the Modern Languages and Literatures. 1919–.

McGraw-Hill Encyclopedia of Science and Technology. 6th ed., 1987. 20 vols. Supplemented by *McGraw-Hill Yearbook of Science and Technology.*

McGraw-Hill Encyclopedia of World Drama. 2nd ed., 1984. 5 vols.

Mythology of All Races. 1916–1932. 13 vols.

New Catholic Encyclopedia. 1967–1979. 17 vols.

New Century Cyclopedia of Names. 1954. 3 vols.

New Encyclopaedia Britannica. Supplemented by *Britannica Book of the Year.* 1938–.

New Grove Dictionary of Music and Musicians. 1980. 20 vols.

Oxford Classical Dictionary. 2nd ed., 1970.

Oxford Companion to American Literature. 5th ed., 1983.

Oxford Companion to English Literature. 5th ed., 1985.

Oxford Companion to Film. 1976.

Oxford History of English Literature. 1945–.

Princeton Encyclopedia of Poetry and Poetics. 1974.

Statesman's Yearbook: Statistical and Historical Annual of the States of the World. 1864–.

Statistical Abstract of the United States. 1878–.

Webster's Biographical Dictionary. 1980.

World Almanac and Book of Facts. 1868–.

The library catalog

The basic tool for finding books in the library is the catalog. It may be in the form of a card catalog, a microform catalog, or a computerized catalog. Books are listed alphabetically by author, title, and subjects with helpful cross-references that will steer you to new aspects of your topics. Reproduced on pp. 340–342 are four typical catalog cards—actually four copies of the same Library of Congress card to be filed in four different ways. Notice the typed title and subject headings.

In some libraries the card catalog has been supplemented or replaced by catalogs on microform. These new catalogs require comparatively little space and can be made available in locations outside the library. A COM (Computer Output Microform) catalog is commonly on microfiche or microfilm and has author, title, and subject entries arranged in the same way as the card catalog. Computerized, or on-line, catalogs are being used by an increasing number of libraries to take advantage of the computer's capabilities to store and retrieve the information formerly found on catalog cards. These systems allow the library user to sit at a terminal and search for material by author, title, subject, or any combination of these. Many on-line catalogs also allow the user to search for library holdings in a variety of new ways, such as by publisher, year of publication, library classification number, or international standard book number (ISBN). Often persons using computer modems at locations away from the library can gain access to these catalogs.

Subject Card

subject (sometimes in red)

Library of Congress
call number

author
title of book

publication
information

QL737
C4M36

CETACEA

McNally, Robert, 1946-
So remorseless a havoc : of dolphins, whales, and men / by
Robert McNally ; illustrations by Pieter Arend Folkens. — 1st
ed. — Boston : Little, Brown, c1981.

pages and technical
data

xvi, 268 p. : ill. ; 24 cm.
Bibliography: p. 253-259.
Includes index.
ISBN 0-316-562920 : $12.95

subject headings
under which this
book is listed

1. Cetacea.   2. Whaling.   I. Title.

QL737.C4M36   1981   333.959'9—dc19   81-5958
                              AACR 2   MARC

Dewey decimal
call number

Library of Congress

Another Subject Card

```
QL737      WHALING
C4M36
           McNally, Robert, 1946-
               So remorseless a havoc : of dolphins, whales, and men / by
           Robert McNally ; illustrations by Pieter Arend Folkens. — 1st
           ed. — Boston : Little, Brown, c1981.

               xvi, 268 p. : ill. ; 24 cm.
               Bibliography: p. 253-259.
               Includes index.
               ISBN 0-316-56292-0 : $12.95


               1. Cetacea.   2. Whaling.    I. Title.
               QL737.C4M36   1981          333.95′9—dc19         81-5958
                                                                 AACR 2   MARC

           Library of Congress
```

Author Card

```
QL737
C4M36
           McNally, Robert, 1946-
               So remorseless a havoc : of dolphins, whales, and men / by
           Robert McNally ; illustrations by Pieter Arend Folkens. — 1st
           ed. — Boston : Little, Brown, c1981.

               xvi, 268 p. : ill. ; 24 cm.
               Bibliography: p. 253-259.
               Includes index.
               ISBN 0-316-56292-0 : $12.95


               1. Cetacea.   2. Whaling.    I. Title.
               QL737.C4M36   1981          333.95′9—dc19         81-5958
                                                                 AACR 2   MARC

           Library of Congress
```

Title Card

```
QL737     So remorseless a havoc
C4M36
          McNally, Robert, 1946-
             So remorseless a havoc : of dolphins, whales, and men / by
          Robert McNally ; illustrations by Pieter Arend Folkens. — 1st
          ed. — Boston : Little, Brown, c1981.

             xvi, 268 p. : ill. ; 24 cm.
             Bibliography: p. 253-259.
             Includes index.
             ISBN 0-316-56292-0 : $12.95


             1. Cetacea.   2. Whaling.   I. Title.
          QL737.C4M36   1981            333.95′9—dc19            81-5958
                                                          AACR 2   MARC

          Library of Congress
```

Periodical indexes

Articles in magazines and newspapers are a rich source of information on almost every subject. Some of the more important periodical indexes are listed below. Some of them are books, some are data bases, some are microform indexes (that is, on microfiche or microfilm), and some are published in more than one form. *The Readers' Guide to Periodical Literature,* for example, is found in many libraries in large green buckram volumes in the Reference Department. It is also available as a database to be used on-line or on compact disc. Your reference librarian can tell you which indexes are available, in what format, and how you can use them.

PERIODICAL INDEXES

Readers' Guide to Periodical Literature. 1900–.
An index to the most widely circulated American periodicals.

Infotrac. 1982–.
Computerized index covering five newspapers and nearly one thousand general-interest periodicals. Especially useful for current topics.

Magazine Index. 1977/78–.
A microform index covering all the periodicals in *Readers' Guide* and many more. Especially useful for current topics.

Nineteenth Century Readers' Guide to Periodical Literature. 1890–1899.
Author and subject index to some fifty English-language general periodicals of the last decade of the nineteenth century.

Poole's Index to Periodical Literature. 1802–1906.
An index by subject to the leading British and American periodicals of the nineteenth century.

Humanities Index. 1974–. Preceded by *Social Sciences and Humanities Index.*

Social Sciences Index. 1974–. Preceded by *Social Sciences and Humanities Index.*

Social Sciences and Humanities Index. 1965–1974. Formerly *International Index.* 1907–1965.
Author and subject index to a selection of scholarly journals.

British Humanities Index. 1962–.
Supersedes in part *Subject Index to Periodicals.* 1915–1922, 1926–1961.
Subject index to British periodicals.

Applied Science and Technology Index. 1958–.
Cumulative subject index to a selection of English and American periodicals in such fields as aeronautics, automation, chemistry, electricity, engineering, physics.

Art Index. 1929–.
   "Cumulative Author and Subject Index to a Selected List of Fine Arts Periodicals."

Biography Index. 1946–.
   "Cumulative Index to Biographical Material in Books and Magazines."

Biological and Agricultural Index. 1964–.
   Continues *Agricultural Index.* 1919–1964. "Cumulative Subject Index to Periodicals in the Fields of Biology, Agriculture, and Related Sciences."

Book Review Digest. 1905–.
   Index to book reviews. Includes excerpts from the reviews.

Book Review Index. 1965–.

Business Index. 1979–.
   A microform index useful for current business topics.

Business Periodicals Index. 1958–.
   Cumulative subject index to periodicals in all fields of business and industry.

Current Index to Journals in Education. 1969–.
   Covers "the core periodical literature in the field of education" and "peripheral literature relating to the field of education."

Education Index. 1929–.
   "Cumulative Subject Index to a Selected List of Educational Periodicals, Proceedings, and Yearbooks."

General Science Index. 1978–.
   Science literature for the nonspecialist.

Index to Black Periodicals. 1950–.
   "An Index to African-American Periodicals of General and Scholarly Interest."

Industrial Arts Index. 1913–1957.

> In 1958 divided into the *Applied Science and Technology Index* and *Business Periodicals Index.* "Subject Index to a Selected List of Engineering, Trade and Business Periodicals."

Music Index. 1949–.

> Index by author and subject to a comprehensive list of music periodicals published throughout the world.

National Newspaper Index. 1979–.

> An on-line index to the *New York Times, Wall Street Journal, Christian Science Monitor, Los Angeles Times, Washington Post,* and other newspapers.

Newsbank Index. 1982–.

> A microfiche index of selected articles in over one hundred city newspapers in the United States.

New York Times Index. 1851–.

> "Master-Key to the News Since 1851."

Public Affairs Information Service. Bulletin. 1915–.

> Subject index to periodicals and government publications chiefly in the social sciences.

Suppose you are writing about the current controversy over dolphins. Looking under **Dolphins** in the *Readers' Guide to Periodical Literature,* November 1987, you find the following entry:

subject heading ⟶ **Dolphins**

cross-reference ⟶ *See also*
     Killer Whales ⟵ *author*

title ⟶ Dolphin talk [interview with D. Reiss] G. Hartwell.

periodical ⟶ *Oceans* 20:62-3 Mr/Ap '87 ⟵ *pages*

volume — Strange pool-fellows [dolphin swim seminar at Sealand Aquarium. West Brewster, Mass.] L.W. Kloss. il *Travel Holiday* 167:8+ Ap '87 ⟵ *date*
     Think like a dolphin [role models for managers learning strategic thinking skills] *Futurist* 21:52 Mr/Ap '87

Consult indexes to periodicals in specific subject areas as well as the more general indexes such as the *Readers' Guide*. Under the heading **Dolphins and porpoises** in the *General Science Index* for November 1987, you find the following information:

> **Dolphins and porpoises**
> Dolphin deaths raise pollution worries. M. Holderness, il *New Sci* 115:22 S 3 '87
> Habitats for cetaceans. *Oceans* 20:57 Jl/Ag '87
> Into the world of orcas. A. Morton. il *Int Wildl* 17:12-17 S/O '87
> Keeping an ear on orcas. D. Hand. il *Oceans* 20:10-19 + Jl/Ag '87
> Sonic punch: dolphins and whales generate "bangs" that may stun prey. T. M. Beardsley. *Sci Am* 257:36 O '87

NOTE: If you do not know the abbreviation for a periodical, check the front matter in the index.

With the information you have gathered from periodical indexes, you should be able to find articles on your subject. Check the serials list or catalog to see whether the periodicals are in your library. You probably will not have time to read carefully all the articles written on your topic; you can exclude some merely by studying their titles in the periodical indexes.

Your **working bibliography** should grow as you proceed. Be sure to include all the information that will help you find each item listed: along with the author and title, you will need the library call number for books and the date, volume, and page number for articles.

43c   Distinguish between primary and secondary sources and evaluate materials.

Primary sources are those about which other materials are written. *Moby-Dick* is a primary source; writings about *Moby-Dick* are

secondary sources. The terms distinguish between which material comes first (primary) and which follows (secondary). Sometimes what was once secondary—say, a book about Puritan New England—becomes primary material when it is the subject of later writings (in this instance, articles about the book). Primary sources for a study of tourists would consist of published and unpublished diaries, journals, and letters by tourists, interviews of tourists, and the like. The writings of journalists and historians about tourists would be secondary sources.

When considering which secondary sources to use in your research paper, you must know when a work was written, what information was available to its author, the reputation and reliability of the author, and, insofar as you can judge, the soundness of the author's argument and use of evidence. *The fact of publication is not a guarantee of truth, wisdom, objectivity, or perception.* Look with a critical eye at all the secondary material you discover in your search and rely only on those sources that are worthy (see p. 264).

43d Take accurate and effective notes from sources.

After you have compiled a working bibliography (see **43b**), located some of the sources you wish to use, and done some preliminary reading, you are ready to begin collecting specific material for your paper. If you are writing a formally documented paper, make a **bibliography card** for each item as you examine it. This will be a full and exact record of bibliographical information, preferably on a three-by-five-inch filing card. From these cards you will later compile your list of Works Cited or References. A sample card is shown on p. 348. The essential information includes the name of the author, the title of the work, the place and date of publication, and the name of the publisher. If the work has an editor or a translator, is in more than one volume, or is part of a series, these facts should be included. For later checking, record the library call number.

Bibliography Card

(reduced facsimile—actual size 3 x 5 inches)

author ———→ Bryden, M.M., and Richard Harrison, eds.

title

Research on Dolphins ←

date of
publication

place of
publication
and name of ———→ Oxford : Clarendon, 1986. ←
publisher

call number ———→ QL 737
C 432 R47

You may find it
convenient to keep
these cards separate
from those on which
you take notes. They
will eventually be
used in making up
the list of Works
Cited in your paper.

CAUTION: Carelessness in jotting down call numbers can lead to wasted time and confusion. Double-check to be sure you have recorded them correctly.

For **note-taking**, your next step, use cards or slips of paper uniform in size. Cards are easier to use than slips because they withstand more handling. Develop the knack of skimming a source so that you can move quickly over irrelevant material and concentrate on pertinent information. Use the table of contents, the section headings, and the index to find chapters or pages of particular use to you. As you read and take notes, consider what subtopics you will use. The two processes work together: your reading will give you ideas for subtopics, and the subtopics will give direction to your note-taking.

At this point you are already in the process of organizing and outlining the paper. Suppose you wish to make a study of current attitudes toward dolphins. You might work up the following list of tentative subjects:

> aspects of dolphin intelligence
> captivity vs. freedom for dolphins
> cooperation between dolphins and humans
> references to dolphins in ancient and modern times
> use of dolphins in military situations
> mysteries surrounding dolphins

These headings may not be final. Always be ready to delete, add, and change headings as you read and take notes. At this stage, it is neither possible nor necessary to determine the final order of headings.

To illustrate the methods of note-taking, suppose you have found the following passage about dolphins kept in captivity.

In our marinelands, dolphins find companions in captivity. They do tricks. They participate in shows. They have fans. But the behavior patterns which they develop in these conditions have very little to do with those which obtain when a dolphin lives at liberty in the sea. We have succeeded in creating a personality common to captive dolphins. And it is that personality that is being studied, without taking into sufficient account that we are dealing with animals that have been spoiled and perverted by man.

> JACQUES-YVES COUSTEAU AND PHILIPPE DIOLÉ
> *Dolphins*

You may make notes on this passage by paraphrasing, by quoting, or by combining short quotations with paraphrasing (see **43f**). The card on p. 351 identifies the source, gives a subject heading, indicates the page number, and then records relevant information in the student's own words. It *extracts* items of information instead of merely recasting the entire passage and line of thought in different words. Notice the careful selection of details and the fact that the paraphrase is considerably shorter than the original. The first card on p. 352 records a direct quotation, and the second combines quotation with paraphrase.

If you cannot determine just what information you wish to extract when you are taking notes, copy the entire passage. For later reference you must be careful to show by quotation marks that it is copied verbatim.

NOTE: The writer inserted a slash [/] in the quotation on the first card on p. 352 to indicate a break between pages.

A note card should contain information from only one source. If you keep your notes on cards with subject headings, you can later arrange all your cards by topics—a practical and an orderly procedure. Remember, the accuracy of your paper depends largely on the accuracy of your notes.

A photocopying machine can guarantee accuracy and save you time. At an early stage in your research it is not always possible to know exactly what information you need. Photocopy

Paraphrased Notes

(reduced facsimile—actual size 3 x 5 inches)

subject heading →

page number →

Dolphins in Captivity

Cousteau and Diolé

158-59 It is true that dolphins may discover much to like in a captive situation, and they appear to adjust well. Nevertheless, those who study them should remember that these animals have undergone a drastic change, and we cannot assume their behavior to be the same as that of dolphins in the open sea.

Identification of source. Full bibliographical information has been taken down on the bibliography card.

Quotation

Dolphins in Captivity

Cousteau and Diolé

158-59 "But the behavior patterns which
they develop in these conditions have very
little to do with those which obtain when a
dolphin lives at liberty in the sea./We have
succeeded in creating a personality common
to <u>captive</u> dolphins" (italics mine).

Quotation and Paraphrase

Dolphins in Captivity

Cousteau and Diolé

158 Dolphins may discover much to like
in a captive situation, and they appear to
adjust well. "But the behavior patterns
which they develop in these conditions have
very little to do with those which obtain
when a dolphin lives at liberty in the sea."

Quotation

Cooperation Between Humans & Dolphins

McNally

58 "Dolphins do, of course, support their own kind in distress, and in rescuing a troubled human they are simply extending that behavior to another species."

Quotation

Legends About Dolphins

McNally

58 "Scientists also used to deride the stories about dolphins carrying shipwrecked and drowning people to safety ashore. Some of these stories are the hallucinations of people who have been pulled back from the threshold of death, but others have the ring of truth or the corroboration of eyewitnesses."

Paraphrase

> *Mysteries of Dolphin Behavior*
>
> McNally
>
> 58-59 Strangely, dolphins do not act kindly toward humans just for the sake of reward. They help human beings in the water in the same manner as they would one of their own, but the question is how they understand that we need aid.

Quotation and Paraphrase

> *Mysteries of Dolphin Behavior*
>
> McNally
>
> 58-59 Dolphins will "cooperate or associate with humans" even when we do not reward them, and we are not sure why this is. Especially mysterious is how they "recognize that a human ... is in trouble" and "that taking this unknown creature ashore" is the correct action.

some of the longest passages, and then you can study them, digest them, and take notes later.

Read the following passage, which deals with the cooperation between dolphins and human beings:

> Sometimes cetaceans cooperate or associate with humans even when there appears to be nothing in it for them. Scientists . . . used to deride the stories about dolphins carrying shipwrecked and drowning people to safety ashore. Some of these stories are the hallucinations of people who have been pulled back from the threshold of death, but others have the ring of truth or the corroboration of eyewitnesses. Dolphins do, of course, support their own kind in distress, and in rescuing a troubled human they are simply extending that behavior to another species. Yet how does the dolphin recognize that a human, a species it rarely encounters, is in trouble, and how does it figure out that taking this unknown creature ashore is the right thing to do?
>
> ROBERT MCNALLY
> *So Remorseless a Havoc: Of Dolphins, Whales and Men*

Now study the note cards on pp. 353–354, which are based on this passage. Observe the variety in subject headings and treatments.

43e Produce an effective outline.

Since a research paper is usually longer and more complex than ordinary college papers, a good outline is likely to be of even greater help in your attempts to organize, expand, and delete materials. Consult **39c** about the various kinds of outlines.

After you have worked out a tentative outline, study it to be sure that you have included all the areas you wish to cover, that you have grouped subjects under the most appropriate headings, and that you have arranged the aspects of your discussion in the order that will produce the maximum effectiveness.

The thought that you put into a full and coherent outline is an excellent investment, which will pay off when you write the paper. You will find that you have worked out many of the problems of organization and made essential connections in preparing the outline that you would otherwise have to face during the composition of the paper. See the model outline on p. 373.

43f Acknowledge your sources; avoid plagiarism. Quote and paraphrase accurately.

Using others' words and ideas as if they were your own is a form of stealing. It is called **plagiarism**. Avoid it by giving full references to sources.

All direct quotations must be set off or placed in quotation marks and acknowledged in your text.

Even when you take only a phrase or single distinctive word from a source, you should enclose it in quotation marks and indicate where you got it. Compare the sources below with the passages that plagiarize them.

SOURCE

> More than any previous explorer, Cook was well prepared to chart his discoveries and fix their locations accurately. He sailed at a time of rapid advances in methods of navigation; his ship was equipped with every available type of scientific instrument; he had the services of professional astronomers; and he himself had a far more sophisticated understanding of astronomy, mathematics, and surveying techniques than most ship captains.
>
> LYNNE WITHEY
> *Voyages of Discovery: Captain Cook and the Exploration of the Pacific*

PASSAGE WITH PLAGIARISM

> Although we often think of the famous Captain Cook as someone who was more of an adventurer than a systematic explorer, he actually was better prepared than any previous explorer to discover and fix new locations accurately. Many improvements were taking place in sailing. For example, his ship was furnished out with every available type of scientific instrument, and Cook himself had a far more sophisticated understanding of navigation than is often realized.

SOURCE

Although Poe's protagonists question their thoughts and feelings as well as their motivations and analyze the amount of harm they have done, they rarely grow in understanding. They lack the detachment necessary to rectify the situation and the will to carry out their decisions once these are made. Solipsistic in every way, they are caught in the stifling world of their own choice.

BETTINA L. KNAPP
Edgar Allan Poe

PASSAGE WITH PLAGIARISM

Poe's main characters do examine themselves, but they rarely grow in understanding. They are not objective enough to change, and they lack the will to carry out their decisions once these are made. They are weird, or solipsistic, in every way and therefore get stifled in their own choice.

These passages with plagiarism do contain some paraphrasing, but they also use some of the words of their sources without employing quotation marks, and neither indicates a source. If the source were acknowledged at the end but no quotation marks were used around words and phrases taken from the source, the writer would still be guilty of plagiarism. The writer of the passage about Poe was not only dishonest but also careless. He or she obviously did not know the meaning of the word *solipsistic* in the source, guessed that it meant something like "weird," and thus composed a sentence that does not make sense. The writer assumed that the reader would not know what *solipsistic* means either and would not have read the source. Thus plagiarism is also insulting to the reader and impractical since it often calls attention to itself.

Good and bad paraphrasing

A good paraphrase expresses the ideas found in the source (for which credit is always given) but not in the same words. It preserves the sense, but not the form, of the original. It does not retain the sentence patterns and merely substitute synonyms for the original words, nor does it retain the original words and merely alter the sentence patterns. It is a genuine restatement. It is briefer than its source.

SOURCE

> Although Poe's protagonists question their thoughts and feelings as well as their motivations and analyze the amount of harm they have done, they rarely grow in understanding. They lack the detachment necessary to rectify the situation and the will to carry out their decisions once these are made. Solipsistic in every way, they are caught in the stifling world of their own choice.

IMPROPER PARAPHRASE

> Even though Poe's heroes ask themselves about their ideas and emotions and also their reasons for acting and figure out how much damage they have left behind, they do not often gain more wisdom. They do not have the impartiality mandatory to alter the condition and the determination to follow through on their resolutions when they are formulated. Self-involved in all manner, they are snared in the smothering realm of their making (Knapp 155).

PROPER PARAPHRASE

> Poe's main characters do engage in self-examination, but they seldom benefit from it. They are too close to their own problems to solve them, too weak to change. Their whole world is within themselves, and they are destroyed by it (Knapp 155).

Frequently a combination of quotation and paraphrase is effective.

SOURCE

> More than any previous explorer, Cook was well prepared to chart his discoveries and fix their locations accurately. He sailed at a time of rapid advances in methods of navigation; his ship was equipped with every available type of scientific instrument; he had the services of professional astronomers; and he himself had a far more sophisticated understanding of astronomy, mathematics, and surveying techniques than most ship captains.

QUOTATION AND PARAPHRASE

One of Captain Cook's distinctions was the care he took in readying himself for his voyages. He "was well prepared to chart his discoveries and fix their locations accurately." He did not sail without good equipment and competent navigators, and "he himself had a far more sophisticated understanding of astronomy, mathematics, and surveying techniques than most sea captains" (Withey 87–88).

PARAPHRASE

One of Captain Cook's distinctions was the care he took in readying himself for his voyages. He did not sail without good equipment and competent navigators. Indeed, he was well versed in the several sciences so important in long-distance sailing (Withey 87–88).

■ Exercise 1

Choose an article from a periodical; avoid articles that are mainly factual or statistical. Then select two passages from one page of the article and make a photocopy (number the passages 1 and 2). In ten to fifteen words, paraphrase the first passage (which should be thirty to fifty words long). Do not state any ideas of your own. Do not use words from the article. In about twenty-five words, paraphrase the second passage (which should be sixty to seventy-five words). This time combine some of your own ideas with the paraphrase of the source. Turn the photocopy in with your paraphrases.

■ Exercise 2

Turn in to your teacher a photocopy of a single page from an article in a periodical. (Do not use the same page that you used for Exercise 1.) Draw circles around three quotations. Submit with the photocopy the following: (a) a sentence that you quote with no writing of your own, (b) a sentence of your own with short quotations in it from the source, and (c) a sentence of your own that introduces a quoted sentence. Write your part, followed by a comma or a colon, and then the complete quoted sentence.

43g Follow an accepted system of documentation.

The forms of documentation vary with fields, periodicals, and publishers. Instructors frequently have preferences that determine what specific system you will need to conform to in your research paper. The following style books and manuals are often used in general studies.

Achtert, Walter S., and Joseph Gibaldi. *The MLA Style Manual.* New York: Modern Language Assn. of America, 1985.

Chicago Manual of Style. 13th ed. Chicago: U of Chicago P, 1982.

Gibaldi, Joseph, and Walter S. Achtert. *MLA Handbook for Writers of Research Papers.* 3rd ed. New York: Modern Language Assn. of America, 1988.

Publication Manual of the American Psychological Association. 3rd ed. Washington: American Psychological Assn., 1983.

Skillin, Marjorie E., Robert M. Gay, et al. *Words into Type.* 3rd ed. Englewood Cliffs: Prentice, 1974.

Turabian, Kate L. *A Manual for Writers of Term Papers, Theses, and Dissertations.* 5th ed. Chicago: U of Chicago P, 1987.

Webster's Standard American Style Manual. Springfield, MA: Merriam-Webster, 1985.

The model paper on pp. 371–419 follows the system of documentation described in *The MLA Style Manual.* This system is widely used in the humanities.

The model paper on pp. 429–451 illustrates the method of documentation presented in the *Publication Manual of the American Psychological Association* (APA), most frequently used in the social sciences.

43h The MLA Style of Documentation

List of works cited

At the end of a research paper, after the text and after the notes (or endnotes), comes a section called "Works Cited." Here you list alphabetically all the sources that you refer to in the text (see the list for the model paper, pp. 415–419). Study the examples below of forms used in MLA documentation for various kinds of sources.

A BOOK BY A SINGLE AUTHOR

Bell, Bernard W. The Afro–American Novel and Its
    Tradition. Amherst: U of Massachusetts P, 1987.

Marcus, Millicent. Italian Film in the Light of
    Neorealism. Princeton: Princeton UP, 1986.

A PAMPHLET

Toops, Connie M. The Alligator: Monarch of the
    Everglades. Homestead, FL: Everglades Natural
    History Association, 1979.

NOTE: A pamphlet is treated as a book. Name the state when the place of publication (Homestead) is not likely to be known. Use U.S. Postal Service abbreviations.

AN EDITED BOOK

Murry, John Middleton, ed. Journal of Katherine
    Mansfield. New York: Knopf, 1952.

A BOOK WITH TWO EDITORS OR AUTHORS

Manley-Casimir, Michael, and Carmen Luke, eds.
    <u>Children and Television: A Challenge for
    Education</u>. New York: Praeger, 1987.

NOTE: On the title page of this book the following cities are listed as places of publication: New York; Westport, Connecticut; and London. In your entry, use only the first city named.

A BOOK WITH MORE THAN THREE EDITORS OR AUTHORS

Kagan, Sharon L., et al., eds. <u>America's Family
    Support Programs</u>. New Haven: Yale UP, 1987.

NOTE: The title page of this book lists four editors: Sharon L. Kagan, Douglas R. Powell, Bernice Weissbourd, and Edward F. Zigler. Use only the first, followed by *et al.* ("and others").

A TRANSLATED BOOK

Palomino, Antonio. <u>Lives of the Eminent Spanish
    Painters and Sculptors</u>. Trans. Nina Ayala
    Mallory. Cambridge: Cambridge UP, 1987.

A MULTIVOLUME WORK

Hendrick, Burton J. <u>The Life and Letters of Walter
    H. Page, 1855 to 1918</u>. 2 vols. Garden City:
    Doubleday, 1927.

A REPUBLISHED BOOK

Langdon, William Chauncy. <u>Everyday Things in
    American Life, 1776–1876</u>. 1941. New York:
    Scribner's, 1969.

NOTE: The first date is that of original publication.

SECOND OR LATER EDITION OF A BOOK

Brooks, Cleanth, and Robert Penn Warren. <u>Modern
    Rhetoric</u>. 2nd ed. New York: Harcourt, 1958.

AN ESSAY IN A COLLECTION OF ESSAYS OR AN ANTHOLOGY

Powell, Douglas R. "Day Care as a Family Support
    System." <u>America's Family Support Programs</u>.
    Ed. Sharon L. Kagan, et al. New Haven: Yale
    UP, 1987. 115–32.

AN INTRODUCTION, FOREWORD, PREFACE, OR AFTERWORD IN A BOOK

Gibson, James. Introduction. <u>The Complete Poems
    of Thomas Hardy</u>. London: Macmillan, 1976.
    xxxv–xxxvi.

Tobias, Andrew. Foreword. <u>Extraordinary Popular
    Delusions and the Madness of Crowds</u>. By
    Charles Mackay. 1841. New York: Harmony,
    1981. xii–xvi.

NOTE: The word is *foreword*, not *forward*. This reference is to a
modern printing of an older work.

MORE THAN ONE WORK BY SAME AUTHOR

Phillips, Derek L. <u>Abandoning Method</u>. San
    Francisco: Jossey, 1973.

———. <u>Toward a Just Social Order</u>. Princeton:
    Princeton UP, 1986.

———. <u>Wittgenstein and Scientific Knowledge: A
    Sociological Perspective</u>. Totowa, NJ: Rowman,
    1977.

NOTE: Multiple works by the same author are listed alphabetically
by title.

SIGNED ARTICLE IN A REFERENCE WORK

Emmet, Dorothy. "Ethics." <u>International Encyclo-
    pedia of Social Sciences</u>. 1968 ed.

UNSIGNED ARTICLE IN A REFERENCE WORK

"Domino." <u>Encyclopedia Americana</u>. 1986 ed.

GOVERNMENT PUBLICATION

United States. Dept. of Commerce. Internal Trade
    Administration. <u>United States Trade: Perform-
    ance in 1985 and Outlook</u>. Washington: GPO,
    1986.

ARTICLE IN A MONTHLY PERIODICAL

Clark, Earl. "Oregon's Covered Bridges." <u>Travel-
    Holiday</u> Jan. 1987: 71-72.

ARTICLE IN A WEEKLY PERIODICAL

Bernstein, Jeremy. "Our Far-Flung Correspondents."
    <u>New Yorker</u> 14 Dec. 1987: 47-105.

ARTICLE IN A QUARTERLY PERIODICAL (with continuous pagination)

Handlin, Oscar. "Libraries and Learning." <u>American
    Scholar</u> 56 (1987): 205-18.

ARTICLE IN A JOURNAL (not with continuous pagination)

Ardoin, John. "A Pride of Prima Donnas." <u>Opera</u>
    <u>Quarterly</u> 5.1 (1987): 58-70.

NOTE: The issue number (1) has been added.

ARTICLE WITH TWO AUTHORS IN A JOURNAL

Gottfries, Nils, and Henrik Horn. "Wage Formation
    and the Persistence of Unemployment."
    <u>Economic Journal</u> 97 (1987): 877-84.

BOOK REVIEW IN A PERIODICAL

Davin, Delia. Rev. of <u>Revolution Postponed: Women</u>
    <u>in Contemporary China</u>, by Margery Wolf.
    <u>Public Affairs</u> 59 (1986-87): 684-85.

SIGNED ARTICLE IN A DAILY NEWSPAPER

Patton, Scott. "Rules of the Ratings Game."
    <u>Washington Post</u> 19 Jan. 1988: D7.

NOTE: "D" refers to the section of the paper; "7," to the page
within that section.

UNSIGNED ARTICLE IN A DAILY NEWSPAPER

"Space Station Is Falling, to Hit Earth in February."
    <u>Orlando Sentinel</u> 5 Jan. 1991: A16.

SIGNED EDITORIAL IN A DAILY NEWSPAPER

Blair, Sharon. "A Charity that Breeds Success."
      Editorial. <u>Wall Street Journal</u> 31 Dec. 1990: 6.

NOTE: No section is indicated before the page number because this particular issue of the newspaper was numbered consecutively throughout.

UNSIGNED EDITORIAL IN A DAILY NEWSPAPER

"The Odds of Ignorance." Editorial. <u>Miami Herald</u>
      19 Jan. 1988: A10.

UNPUBLISHED DISSERTATION

Kenney, Catherine McGehee. "The World of James
      Thurber: An Anatomy of Confusion." Diss.
      Loyola of Chicago, 1974.

ABSTRACT IN *DISSERTATION ABSTRACTS INTERNATIONAL*

Kenney, Catherine McGehee. "The World of James
      Thurber: An Anatomy of Confusion." <u>DAI</u> 35
      (1974): 2276A. Loyola of Chicago.

A FILM

Sturgis, Preston, dir. <u>The Great Moment</u>. With Joel
      McCrea and Betty Field. Universal, 1944.

A TELEVISION PROGRAM

God Bless America and Poland, Too.   Writ. Marian
      Marzyski.   The American Experience.   PBS.
      WMFE, Orlando.   10 Jan. 1991.

A VIDEOTAPE

The Remarkable Bandicoots.   Videocassette.   Dir. David
      Moore.   Pacific Arts Video, 1985.   50 min.

Parenthetical references

The list of Works Cited indicates clearly what sources you used,
but precisely what you derived from each entry must also be
revealed at particular places in the paper. Document each idea,
paraphrase, or quotation by indicating the author (or the title if
the work is anonymous) and the page reference at the appropriate
place in your text.

> The extent of dissension that year in Parliament has been pointed
> out before (Levenson 127). *[Writer cites author and gives author's
> name and page number in parentheses.]*

> Accordingly, no "parliamentary session was without severe dissen-
> sion" (Levenson 127). *[Writer quotes author and gives author's name
> and page number in parentheses.]*

> Levenson points out that "no parliamentary session was without
> severe dissension" (127). *[Writer names author, places quotation
> marks at beginning of quotation and before parenthesis, and gives
> only page number in parentheses.]*

These references indicate that the quotation is to be found in the
work by Levenson listed in Works Cited at the end of the text of
the paper.

   If two authors in your sources have the same last name, give
first names in your references.

It has been suggested that the general's brother did not arrive until the following year (Frederick Johnson 235).

The authenticity of the document, however, has been questioned (Edwin Johnson 15).

If two or more works by the same author are listed in Works Cited, indicate which one you are referring to by giving a short title: (Levenson, *Battles* 131). The full title listed in Works Cited is *Battles in British Parliament, 1720–1721*.

If a work cited consists of more than one volume, give the volume number as well as the page: (Hoagland 2:173–74).

Some sources—the Bible and well-known plays, for example—are cited in the text of the paper but not listed as sources in Works Cited. Use the following forms:

. . . the soliloquy (*Hamlet* 2.2). [The numbers designate act and scene. Some instructors prefer roman numerals: *Hamlet* II.ii.]

. . . the passage (1 Kings 4.3). [That is, chapter 4, verse 3.]

For further illustrations of parenthetical references to works cited, see the model research paper that follows the MLA method.

Notes (endnotes, footnotes)

You will probably need few numbered notes (called "footnotes" when at the bottom of the page, "notes" at the end of the paper) because sources are referred to in the text itself and listed after the body of the paper. Notes are used to explain further or to comment on something you have written. To include incidental information in the text itself would be to interrupt the flow of the argument or to assign undue importance to matters that add to the substance only tangentially. Make your decision as to where to place such information—in the text or in a note—on the basis of its impact and direct relevance.

Notes are also useful in referring to sources other than those mentioned in the text or in commenting on sources. If you wish to list several books or articles in connection with a point you are making, it may prove awkward to include all of this information in parentheses. Therefore, a note is preferable.

The sign of a note is an Arabic numeral raised slightly above the line at the appropriate place in the text. The *MLA Handbook* recommends that notes be grouped at the end of the paper. Notes should be numbered consecutively throughout.

EXAMPLES

[1] See also Drummond, Stein, Van Patten, Southworth, and Langhorne.
[Lists further sources. Full bibliographic information given in Works Cited at end of paper.]

[2] After spending seventeen years in Europe, he returned to America with a new attitude toward slavery, which he had defended earlier.
[Remark parenthetical to main argument but important enough to include in a note.]

[3] This biography, once considered standard, is shown to be unreliable by recent discoveries.
[Evaluates a source.]

[4] Detailed census records are available only for the 1850–1880 period. Earlier censuses do not provide the exact names of inhabitants nor any financial or social information. Later census schedules that do have detailed information on individuals are not available to the public; one may obtain only summaries of the data gathered.
[Provides background on a source.]

[5] Andrews uses the term *koan* to mean any riddle. To prevent confusion, however, *koan* will be used in this paper to designate only riddles in the form of paradoxes employed in Zen Buddhism as aids to meditation.
[Clarifies terminology.]

For further illustrations of notes, see the model research paper, pp. 407–413.

43i   Model research paper, MLA style

A model research paper in the MLA style of documentation, with an outline and accompanying explanations, is given on the following pages.

GENERAL APPEARANCE AND MECHANICS

> Allow ample and even margins.
> Indent five spaces for paragraphs.
> Leave two spaces after periods and other terminal punctuation.
> Leave one space after other marks of punctuation.
> Double-space between lines in the text, the notes, and entries in the Works Cited section.
> Set off a quotation of five or more typed lines. Begin a new line, indent ten spaces from the margin, do not add quotation marks, and double-space unless your instructor specifies single spacing. If you quote only one paragraph, do not indent the first line.

Compose a title page for your research paper if your instructor requires it. (MLA does not require it.) Balance the material on the page. Center the title and place it about one-third of the way down from the top of the page. Include your name and the name and section number of the course as indicated on the opposite page, or follow the specific preferences of your instructor.

The Dolphin Controversy

By Curtis Washington

English 101

Section 3

If your instructor requests that you submit an outline with your paper, it should occupy a separate, unnumbered page following the title page and should follow the form for the outline illustrated on pp. 232, 373.

If your instructor requests that you include a thesis statement as part of your outline, place it between the title and the first line of the outline.

The Dolphin Controversy

I. Appeal of dolphins

    A. In modern times

    B. In ancient times

II. Dolphin as center of controversy

    A. Recentness of debate

    B. Nature of debate

III. Question of dolphin intelligence

    A. Dolphins as equals to humans

    B. Dolphins as inferior to humans

IV. Question of keeping dolphins in captivity

    A. U.S. Navy's training of dolphins

    B. Use of dolphins in Persian Gulf

    C. Arguments for and against such use

V. Emotional ingredients in controversy

    A. Dolphins as evokers of love, sympathy, mystery

    B. Dolphins as benevolent superbeings

    C. Attitudes of humans as reflections of their own deep needs

Place page numbers in the upper right-hand corner, two lines above the first line of text. Use Arabic numerals; do not put a period after the number. (According to MLA style, you may add your last name before the number on each page.)

Center the title on the page. Double-space to the first line of the text.

The first and second paragraphs make up the introductory section of this paper. The writer first establishes the current popularity of dolphins and then illustrates that dolphins have long captured the human imagination. The reader's interest is thus stimulated by the writer's calling attention to this long tradition of human involvement with dolphins.

Flipper is underlined because it is the title of a television series. Titles of individual episodes within a series are placed in quotation marks rather than underlined.

The parenthetical reference to Glueck illustrates the most frequently used form of documentation: author's last name, no punctuation, page number.

Washington 1

The Dolphin Controversy

Perhaps no creature is currently more widely revered than the dolphin. Its exuberance, intelligence, and benevolence have captured the modern imagination. Its image is to be seen on restaurant advertisements and book covers and on motel logos. Its form has been imitated by the manufacturers of children's stuffed toys and by the makers of plastic floats for use in water. A professional football team is called the Dolphins, and for many years young people have thrilled to the adventures of a dolphin in the television series Flipper (now in reruns).

The popularity of dolphins is not, however, strictly a modern phenomenon. In ancient times, "the dolphin alone or in association with male and female deities occurred frequently in sculptural or mosaic or painted form on the mainlands and islands of the entire Mediterranean" (Glueck 349). Classical Greek and Roman literatures abound with references to dolphins and their interaction with human beings. In his Lives the first-century biographer Plutarch tells how a

The reference to "Dolphin" cites an unsigned article in an encyclopedia. No page number is needed for references to entries in encyclopedias.

The sentence beginning with "Delphi" illustrates the technique of combining one's own words with quoted words.

When a source has coauthors, cite the last names of both. Accent marks (as in Diolé) should be carefully reproduced.

Place note numbers slightly above the line of type and after marks of punctuation. Do not leave a space before the number; do not place a period after the number. Number notes consecutively throughout the paper.

Delphis is not documented; the translation or definition of words requires no documentation.

dolphin saved Odysseus' son, Telemachus, from drowning. Long believed to be legends, such stories now
appear to rest "in part at least, on factual grounds"
("Dolphin"). Delphi, "the most famous sanctuary of
Greece" and, according to ancient belief, "the center
of the world," was given its name—if a long-standing
explanation is to be believed—because it was thought
that the god Apollo, who supposedly dwelt in that
place, "first appeared there in the form of a dolphin"
(Cousteau and Diolé 237).[1] The Greek word for dolphin
is <u>delphis</u>. In 330 B.C. Aristotle wrote with relative
accuracy of dolphins in his <u>History of Animals</u>, and
Pliny the Elder, who died in A.D. 79, related in his
<u>Natural History</u> the story of a dolphin that "would
daily carry a boy on his back across the Bay of Baiae
to Puteoli, which is near Naples" (Glueck 359).

Human beings have been fascinated with dolphins
for so long that some scientists, anthropologists, and
sociologists feel that the current (or "New Age") interest is merely a continuation of past attitudes.
According to Robert McNally, "there is little new in

In the third paragraph the writer states the thesis: that despite a continuing fascination with dolphins, human beings have recently begun to quarrel in earnest about them—a new phenomenon.

Within the quotation brackets have been used to explain the term *cetaceans* since this definition is needed and is not a part of McNally's statement. The writer has added an interpolation, for which brackets rather than parentheses are used. Since the author of this quotation has already been identified ("According to Robert McNally"), only a page number is needed in parentheses after the quotation.

Underlining should be done sparingly. Here the writer underlines *has* because of the emphasis he wishes to place on that word within this centrally important sentence.

At the end of this paragraph, the writer clearly indicates that he will be discussing three specific questions that are basic to the modern controversy centering on dolphins. By stating these questions succinctly, he thus gives a strong direction to his essay. Readers know precisely where they will be going.

In the paragraph beginning with "Scientists know," the writer begins to address the problem of understanding dolphins' makeup, emphasizing that mystery still prevails in many areas.

the New Age view of the cetaceans [the general term
for the scientific order that includes whales, porpoises, and dolphins]. It is only the anthropocen-
trism of the centuries with a contemporary twist"
(231). Despite this opinion, something new and highly
significant has become a part of the modern concern
with dolphins: heated controversy. Dolphins have been
popular with humankind for centuries, but never before
have they been the center of intense debate. Ironically, opinions about these kindly mammals long
thought to be our guardian angels of the deep are now
pitting humans against each other with startling intensity. Though the dolphin controversy is many
sided, three aspects of it, which can be expressed as
questions, are especially prominent: (1) What is a
dolphin? (2) Should dolphins be kept in captivity?
(3) Is it immoral to use dolphins for military purposes?

　Scientists know much more than ever before about
dolphins--about the large size of their brains, their
ability to use their splendid sonarlike faculty for

Single quotation marks are used to show quoted words within a quotation.

The parenthetical references to "What is Killing?" and "Dolphin Die-off" cite unsigned articles in weekly magazines. No page numbers are needed in these instances because the articles are only one page long each. Note that shortened versions of the titles are used.

The writer now turns to the most important aspect of this issue of dolphin nature: intelligence.

"echolocation," their recently discovered weapon "in the form of a sound beam . . . to debilitate or 'stun' their prey" (Morris 394).[2] Some researchers now hypothesize that dolphins have a "magnetic receptor system" (Klinowska 401), which enables them to find their way over long distances.[3] Yet each discovery appears to raise new questions and to fuel debate. The more we learn about dolphins, the greater the mystery. For example, dead dolphins have recently washed ashore along the Atlantic Coast in large numbers. Investigators have undertaken to explain why they are dying. Some "scientists suspect a viral or bacterial agent" ("What Is Killing?"). Others argue vehemently that pollution is the culprit. Like so much else having to do with this fascinating mammal, comments one reporter, "the cause of the dolphin deaths may remain a mystery as deep as the sea itself" ("Dolphin Die-off").

Scientists and others avidly interested in dolphins can disagree in friendly fashion over what is causing the deplorable "die-off," as it is called.

In introducing the quotation from John C. Lilly, the writer gives the date of the author's book, 1961, not because such information in this position is always necessary or even advisable but because in this particular instance he wishes to show something of the chronology of the debate.

Prose quotations of five lines or more should be set off ten spaces as indicated here. Leave the right margin unchanged. MLA style calls for double-spacing of quotations that are set off, the practice followed throughout this model paper. (See pp. 149–150.) Do not indent if only one paragraph (or part of it) is quoted; indent for all paragraphs if more than one is quoted. Do not use quotation marks around quotations that are set off.

When ellipsis points (. . .) are used to indicate omitted words at the end of a sentence, end the sentence with a period, and then add three spaced ellipsis points (. . . .).

The name of the author of the quotation that is set off has been given before the quotation; therefore, it should not be mentioned again in parentheses after the quotation. The title of the work quoted is ordinarily not needed since that information is included in the list of Works Cited at the end of the paper. It is given here because two books by the same author are referred to in the paper, and thus it is needed for clarification.

But they are less coolly objective when another and
much larger issue is raised—the intelligence of dolphins. This subject is at the heart of the question
about what the dolphin is, what its place is in the
hierarchy of nature. After extensive exploration of
their intelligence, John C. Lilly boldly made the
following statement in 1961:

> Eventually it may be possible for humans to
> speak with another species. I have come to
> this conclusion after careful consideration
> of evidence gained through my research ex-
> periments with dolphins. . . . We must strip
> ourselves, as far as possible, of our pre-
> conceptions about the relative place of <u>homo
> sapiens</u> in the scheme of nature. . . . If we
> are to seek communication with other spe-
> cies, we must first grant the possibility
> that some other species may have a potential
> (or even realized) intellectual development
> comparable to our own. (<u>Man and Dolphin</u>
> 17–18)

Both the title of the work cited and the names of its authors are given before the quotation beginning with "attacking," so only the page number is needed in parentheses. The small Roman numeral *x* indicates that the quotation is taken from the front matter, in this instance the preface.

In the quotation beginning with "the dolphin brain," the word *could* is underlined because it appears in italic type in the original. If it had been printed in regular type and the writer of this paper had wished to underline it, he could do so, but he would have had to add in parentheses after the quotation "italics mine" or "emphasis added."

Enough information about the motion picture *The Day of the Dolphin* is given here so that no parenthetical documentation is needed. Note that the title of a motion picture is underlined.

Lilly's determination to place dolphins on an equal footing with human beings and his attempts to break the language barrier between species brought forth a host of supporters and an army of detractors. Karl-Erik Fichtelius and Sverre Sjölander aligned themselves with Lilly's point of view in their book, Smarter Than Man? They admit openly to "attacking the deeply rooted notion that man is the most remarkable thing God ever created" (x). Their "surprising conclusion" is that "the dolphin brain could be superior to ours" (40). These spokesmen for the dolphin's high intelligence attracted such attention that in 1973 Joseph E. Levine produced a motion picture based upon a French novel by Robert Merle called The Day of the Dolphin, starring George C. Scott as a Lilly-like scientist devoted to teaching dolphins the English language and marveling at their intelligence, capacity for love, and unswerving loyalty.

Controversy about Lilly and his followers continues to the present; name calling on both sides is not rare. Robert McNally has concluded that Lilly is

Whenever possible, consult directly the authorities that you quote or cite. In this instance, Curtis Washington did not have access to the writings of René-Guy Busnel but wished to cite his idea. He did so by giving the title of the book in which he found Busnel's views discussed. Since Busnel's actual words are not quoted, the parenthetical entry reads "cited in" rather than "quoted in."

Underlining for emphasis should be used sparingly, but since the writer has not made use of it often and since he is summarizing an important part of the argument, it is effective to underline *is*.

a "poor authority" (230), and Lilly has accused other scientists of retreating in arrogance and cowardice from what he terms "interlock research" with dolphins (Programming 95). Defenders of Lilly and his views claim that "his contributions to our knowledge of cetaceans are legion, and his work has probably saved millions of animals by making the public aware of their intelligence and gentle sensitivity" (Gormley 177-78). That "gentle sensitivity," however, is more imagined than real, according to René-Guy Busnel, who insists that dolphins are no more sensitive and no more capable of feeling loyalty and affection than other animals--perhaps less than some (cited in Cousteau and Diolé 88). Recently Dudok van Heel has written that although many scientists wish to place dolphins in an "elevated niche in the Animal Kingdom," others "maintain that there are not enough arguments to assume that man is not alone in attaining a level of reason, intelligence and emotions, which seem to have made this species unique so far" (163).[4] What, then, is a dolphin--a creature equal or even superior

In the paragraph beginning with "As the argument," the writer turns to the second phase of the dolphin controversy—the question of keeping dolphins in captivity. The organization of the paper is thus tightly controlled; the writer is on course; the reader in tow.

Again, the writer desires to use an important quotation to which he does not have direct access. He indicates where he found the quotation. Such a practice is acceptable only if rarely employed.

to human beings in intelligence or merely a large-brained mammal about which there has been much romanticizing? The question still is unanswered as the adversaries remain locked in an intellectual wrestling match, the dolphins looking on with that playful but enigmatic smile.[5]

As the argument over the intelligence and sensitivity of dolphins has raged, a parallel debate has been increasing in intensity on the question of keeping them in captivity. At one point, John Lilly released all the dolphins that he had been working with because, he said, "I no longer wanted to run a concentration camp for my friends" (quoted in Gormley 178).[6] Jacques Cousteau and Philippe Diolé expressed strong objections to keeping dolphins in aquariums and marinelands (158–59), and Gerard Gormley has referred to captive dolphins as having to serve "life sentences in oceanaria" (40). In 1983 a global conference on "Whales Alive" was held in Boston; it "clearly showed to what extent opinions are divided on the ethics of keeping cetaceans in captivity" (Dudok van Heel 163).

Personal interviews with important and knowledgeable people can be useful secondary sources. Such interviews should be set up in advance and should be conducted with due consideration to the interviewee's time. The interviewer should come well prepared with a list of specific questions. If any part of the interview is to be published, permission of the person being interviewed should be sought. Since all pertinent information documenting the interview is given before the quotation and in the list of Works Cited, no parenthetical entry is needed.

Demands for the rights of dolphins were reminiscent of rallying cries in the civil-rights movement of some years ago. A popular weekly magazine commented that "dolphins have such high intelligence that several animal-rights groups campaign against their confinement in any conditions, likening it to jailing an innocent person" ("Dolphin Die-off").

On the other hand, many researchers and trainers of dolphins make a strong case for such facilities as Marineland of Florida. Michael Bright has written that "dolphinaria provided the first and generally the best opportunity of observing dolphin behavior. Marineland of Florida, for instance, pioneered the early work on the sounds that dolphins make for communication and echolocation" (36). In an interview with Gregory Pyle, assistant director of shows at Marineland of Florida, I raised the question of locking up dolphins to serve "life sentences." Mr. Pyle strongly defended the usefulness of oceanaria, though he is critical of facilities where making money has overshadowed all other activities. To close such

The parenthetical reference to Linden is to a signed article in a weekly magazine. No page number is needed since the article covers only a single page.

attractions as Marineland, he stated, would be to deprive thousands of people of the "aesthetic pleasure derived from seeing dolphins"; it would deprive children especially of the "sheer joy of watching dolphins from close up."

Many who feel that the objections of conservationists and some scientists to dolphins' being retained in cramped pools are misdirected point out that once dolphins are habituated to captivity, they seem to like it so well that they often refuse to leave even if given the opportunity. If forced to leave, they frequently cannot survive in the open sea. That view was generally accepted until recently, when two dolphins that had been in captivity for seven years were freed off the coast of Georgia. By gradually "untraining" them, their liberators proved that "civilized" dolphins can be turned back to the wild state successfully (Linden). Still, whether marinelands and aquariums should all be closed and all scientists be required to study dolphins in the wild state continues to be hotly debated, with emotions running high on both sides.

Beginning with "No issue concerning dolphins," the writer now introduces the third and final phase of the dolphin controversy: the question of training dolphins for use in military situations.

Set off dialogue in a motion picture or a play and use quotation marks as indicated here. If the writer were quoting part of a scene from a play, he would need to identify the act and scene (I.iii). Since motion pictures do not designate acts and scenes, no further documentation is required here.

No issue concerning dolphins, however, is more explosive than that of their use for military purposes. Most of the fireworks appear on one side, for seldom do spokesmen for the military come forth with any kind of justification for their training of dolphins or with information about their experiments involving them. Nongovernment scientists generally try to steer clear of the military. The overall situation is fairly accurately presented in a scene from the motion picture The Day of the Dolphin, in which a scientist, Dr. Jacob Terrell, responds to questions after presenting a brief talk on dolphins. The dialogue is as follows:

> Woman in audience: "What about the experiments that the military--"
>
> Dr. Terrell (interrupts): "I don't know anything about the military."
>
> Woman: "But surely you have heard about the misuse of animals like dolphins that the government is rumored to--"

The writer has added the word *dolphins* to the quotation for clarity. Such additions are always enclosed in brackets. Note also that to make the sentence grammatical in the quotation set off on the next page, the word *training* has been added.

Washington 12

Dr. Terrell: "Just a moment, please. I am
not a political scientist. My degrees are
in biology and zoology and behavioral psychology. The government and I pay very
little attention to each other."

Cousteau and Diolé have commented that "the Navy
is reluctant to have it known that these animals
[dolphins] are being trained for military service"
(75–76). Although the United States government has
denied that this training involves any risk to dolphins, the notion persists that the animals may con-
stitute a sort of modern-day kamikaze squad to be sent
on suicide missions to blow up enemy ships. Michael
Bright's statement is representative of this widespread suspicion:

It is generally known that naval authorities
of several countries are currently [train-
ing], or have in the past trained, dolphins
to fix explosives to the hulls of enemy
vessels, to discover or destroy enemy mines,
or to help detect the presence of enemy

After a quotation that is not set off, place the period *after* the parenthesis. When the quotation is set off, the period comes *before* the parenthesis.

After a quotation that has been set off, you do not necessarily have to begin a new paragraph. Here the same paragraph continues, and "Bright" is not indented.

The parenthetical page reference reads (53, 55), not (53–55), because the passage quoted does not run consecutively but occurs on the two pages indicated, with illustrations occupying the intervening page.

The two words quoted, "Public outcry," do not need to be documented because the reader can clearly see that they are taken from the quotation above, the source of which has already been given.

The writer found this information about the military's killing of whales in several sources; therefore, it can be considered common knowledge for which no specific source is required.

Washington 13

submarines. If they can be trained to do

that, it is not a far stretch of the imagi-

nation to consider them in front-line at-

tacks, perhaps exploding underwater mines--

and killing themselves in the process. (53)

Bright goes on to comment on other possible uses of

dolphins in war but remarks that little is actually

known: "Since for each nation military research is a

top-secret operation, there is little information

available." He concludes that "public outcry has

certainly deterred the use of such peaceful animals

for naval warfare, but it may be that experiments

continue, only now with still greater secrecy" (53,

55).

"Public outcry" indeed there has been, especially

since it has come to light that American military

airplanes used whales for target practice in the late

1940s and obliged Icelandic fishermen some years later

by slaughtering killer whales. Charges of cruelty and

insensitivity, however, have not prevented the Navy

from continuing its work with dolphins. On October

In quotations that are run in with the text (that is, not set off), use single quotation marks to designate a quotation within the quotation.

Note that the tone of the discussion remains essentially objective as the writer makes his way through the phases of the controversy. When dealing with highly volatile current issues, the temptation may be strong to express one's own feelings with excessive zeal. In doing so, however, a writer runs the risk of alienating the audience, which Curtis Washington has deemed to be general readers rather than special interest groups. If he were composing for any of those groups, the tone might be substantially different. In any case, a careful consideration of audience is necessary for an effective essay.

24, 1987, the New York Times reported that "dolphins
apparently trained to search for mines are the latest
addition to American forces in the Persian Gulf. The
Navy acknowledged Thursday that it had sent five dolphins at the request of the American commander in the
Gulf, 'to provide an underwater surveillance and de-
tection capacity'" ("Dolphins Hunt Gulf Mines"). As
usual, few details were available. The Pentagon refused to answer any questions about the operation but
stated that the Navy "has never trained nor does it
intend to train marine animals to perform a task which
could result in intentional injury or death to the
animal."

Such pronouncements have failed to quiet the
controversy, for even if dolphins were not trained for
suicidal missions, there would still be widespread
opposition to their being used to further military
objectives of any kind. Yet if dolphins are indeed
anxious to help preserve humankind from the sharks of
the deep, as countless stories attest, then why not
allow them to protect us from other forms of sharks—

The final paragraph brings the writer full circle. He begins with the phenomenal appeal dolphins have for humans, and he ends with the same idea. For his conclusion, however, he has added an important and thoughtful idea, that we reach out to dolphins and sometimes even think of them as benevolent aliens because of our loss of faith in our own kind. He ends the paper with a strong punch and at the same time offers a viable reason, in his view, for the intensity of the controversy.

Writers often summarize and draw together in their conclusions the material that has been discussed in the paper. Since the three aspects of the controversy have been so clearly put forth here, it would be redundant simply to describe them again at the end.

our nation's enemies?[7] So goes the argument in favor of enlisting our friends the dolphins for duty in the military. They may be our friends, but even so, they have unwittingly stirred up considerable trouble among us.

The reason for this trouble is not merely scientific disagreements. The controversy over dolphins is much more emotional than usual issues dealing with animals because these particular mammals have a special appeal. Indeed, it is difficult for some people to think of them as mere animals at all. They evoke not only sympathy, love, and admiration but also a deep sense of mystery. Michael Bright reports that "some claim dolphins have ESP and can telepathically understand a man's thoughts and moods" (6). Because they are clearly intelligent and, unlike human beings, essentially nonaggressive, qualities (like ESP) are often attributed to them without adequate proof. It is no wonder, then, that widespread outrage develops when news of their victimization in any form is reported.[8] As experiments increasingly reveal startling

new details about their abilities, they take on the aura of benevolent extraterrestrial beings with whom we should communicate for the salvation of the human race. Gradually, Flipper seems to have merged with ET. Dolphins have become our latest heroes partly because we no longer seem to be able to find any heroes among our own kind.

The word *Notes* is centered on the page. Double-space throughout unless instructed otherwise.

Indent the first line of every note five spaces; do not indent succeeding lines. Note numbers are raised slightly above the line. Leave a space between the number and the first word of the note.

As note 1 illustrates, bibliographic references in notes are identical in form to parenthetical documentation in the body of the paper. This note gives further information about ancient associations with the dolphin. One of the primary functions of notes is to include details or arguments that are germane but that would somewhat clutter the text were they presented there. Notes assist in preserving a tight structure in the body of the paper while making it possible to include materials that add further interest or evidence.

Washington 17

Notes

[1] From early times the dolphin was associated with the supernatural. Glueck points out that for some ancient people "the dolphin symbol . . . became an attribute of their chief goddess, standing for succor in peril, safety in danger, security and promise of blessing in the unknown and hereafter" (353). Later, the dolphin became in some circles the symbol of Christ and of rebirth (Cousteau and Diolé 247).

[2] Although these sounds "cannot be heard by human beings, apparently they can be _felt_, at least under certain circumstances," in the form of a light tingle (Ellis 67).

[3] "My conclusion," states Klinowska, "is that these animals are certainly using geomagnetic topography as a base map and may also be using local geomagnetic time cues to provide information about their relative position on that map" (402). Thus an answer to why dolphins mysteriously strand themselves on beaches from time to time—an enigma that two thousand years ago Aristotle puzzled over—has been offered in

Note 5 contains general information ("common knowledge"), which requires no documentation.

The writer found the information given in note 6 not in Lilly's own writings but in a book by another author, which is cited in the parenthetical reference. Though what is reported here may be well known, the writer found it only in a single source and could not assume, therefore, that it is common knowledge, for which documentation would not be needed. The passage from Gormley is given below so that it may be compared with the *paraphrase* of it in note 6.

Within a few weeks after he stopped his research, five of the eight dolphins at his laboratory apparently took their own lives. Some starved themselves; others simply stopped breathing. Dr. Lilly freed the remaining three animals and closed his laboratory, saying he would never again experiment with dolphins unless the animals could come and go as they pleased, working with him when they wished and returning to the sea when they did not. Dr. Lilly devoted the next several years to continued isolation tank and LSD experiments, then in 1975 began another attempt to communicate with dolphins, using a computer language he called Janus. . . . Unless conditions have changed since this writing, his dolphins are captives. (178)

Washington 18

terms that satisfy some scientists: the dolphins "make
mistakes" in the use of their magnetic sense, "which
result in the accidents we know as live strandings"
(427).

4 Despite all the postmortem examinations of
dolphin brains and extensive experimentation with live
dolphins, some prominent scientists feel that the
reason for the large brain size is still to be determined. "An explanation as to why the dolphin brain is
so large, and to whether its brain is functionally
primitive must await more studies of dolphin anatomy,
physiology, and behaviour" (Ridgway 68).

5 The dolphins seen in such attractions as Marineland and Sea World are almost always of the bottlenose variety, which has a fixed expression that
resembles a kindly but mischievous smile.

6 After Lilly had let his last three dolphins
go--five that he had ceased working with a short time
earlier seemed to have willed their own deaths and
died--he shut down his laboratory and declared that he
would experiment with dolphins in the future only if

The quotation in note 7 is from a personal letter, not a published work. If time allows, it is often practical and enlightening to carry your research outside the library. Mr. Gormley kindly responded in a full and interesting letter to a written inquiry, and he granted permission to be quoted. No parenthetical documentation is needed here since the writer and the source of the quotation have already been identified. The entry in the list of Works Cited gives the date of the letter.

they were allowed to be free to visit him and return

to the sea as they wished. After an interval during

which he conducted experiments with LSD and with iso-

lation tanks, he resumed his work with dolphins, now

attempting to break through the language barrier by

utilizing computers, but he was compelled to go

against his earlier statements and to keep his dol-

phins in captivity (Gormley 178).

[7] Gerard Gormley, author of <u>A Dolphin Summer</u>,

indicated in a letter that he is opposed generally to

the exploitation of animals but took a position on the

use of dolphins in military situations that is both

sensible and appealing:

> Given the apparent inevitability of war, I
>
> can go along with the noncombatant use of
>
> dolphins to detect mines and save downed
>
> airmen or sailors cast adrift. This may
>
> risk dolphin lives, but we do that every
>
> time we buy "light meat" tuna or fish caught
>
> in drift nets. Besides, each trained dol-
>
> phin represents such a major investment that

military people aren't likely to take unnecessary risks with them. Still, if dolphins are to be used for noble causes, let
their deeds and sacrifices be made public.
This would elevate them in everyone's eyes,
and might help with better treatment for
their kind in the wild.

8 The very title of Farley Mowat's book Sea of
Slaughter suggests the extent of the author's emotional involvement with these victims of humankind.

Begin on a new page for the list of Works Cited. Center the title. Double-space throughout unless instructed otherwise.

Do not indent the first line of an entry; indent succeeding lines five spaces.

List only those sources actually used in your paper and referred to in parenthetical documentation.

Authors are listed with surnames first. If a book has more than one author, the names of authors after the first one are put in normal order (see the second and third entries).

List entries alphabetically by the authors' last name. If a name is not known, arrange the entry by the first word of the title, excluding *A, An,* and *The* (note that the entry for *The Day of the Dolphin* comes in between "Cousteau" and "Dolphin" rather than later on in the list under *T*).

Give the inclusive pages for articles.

Notice that the important divisions of entries are separated by periods.

The first entry illustrates the standard form for a book.

The entry for Cousteau and Diolé shows the proper method to list a book (a) that has more than one author, (b) that has been translated, and (c) that is part of a series (in this case an unnumbered series).

Entries for films should include the title, director, main actors, distributor, and date of release.

The entry for "Dolphin" illustrates the proper way to list an unsigned article in an encyclopedia.

Unsigned articles in weekly periodicals should be listed as indicated by the entries for "The Dolphin Die-off" and "What Is Killing the Atlantic's Dolphins?"

Be careful with foreign words and names. "Dudok van Heel" is the surname of the author, not "Heel" or "van Heel." This entry illustrates the proper way to list an essay in a collection.

For the proper form in listing an article in a monthly periodical, see the entry for Ellis. An entry for an article in a quarterly periodical is shown on p. 364.

Washington 21

Works Cited

Bright, Michael. <u>Dolphins</u>. New York: Gallery, 1985.

Bryden, M. M., and Richard Harrison, eds. <u>Research on
Dolphins</u>. Oxford: Clarendon, 1986.

Cousteau, Jacques-Yves, and Philippe Diolé. <u>Dolphins</u>.
Trans. J. F. Bernard. The Undersea Discoveries
of Jacques-Yves Cousteau. Garden City: Doubleday, 1975.

<u>The Day of the Dolphin</u>. Dir. Mike Nichols. With
George C. Scott. Twentieth Century-Fox, 1973.

"Dolphin." <u>Encyclopedia Americana</u>, 1986 ed.

"The Dolphin Die-off." <u>U.S. News & World Report</u>
24 Aug. 1987: 12.

"Dolphins Hunt Gulf Mines." <u>New York Times</u> 24 Oct.
1987: 3.

Dudok van Heel, W. H. "From the Ocean to the Pool."
<u>Research on Dolphins</u>. Ed. M. M. Bryden and
Richard Harrison. Oxford: Clarendon, 1986.
163-82.

Ellis, Richard. "Dolphins: The Mammal Behind the
Myth." <u>Science Digest</u> Jan. 1982: 62-67.

Carefully reproduce all markings, especially in foreign words, as illustrated in the names of the authors in the entry for Fichtelius.

The second entry for Gormley shows how to list a personal letter.

When more than one work by the same author is listed, do not repeat the name but type three unspaced hyphens followed by a period in place of the author's name in entries after the first. (See the entry that follows that for Lilly's book *Man and Dolphin*.)

The entry for Linden shows the proper way to list a signed article in a weekly magazine.

Washington 22

Fichtelius, Karl-Erik, and Sverre Sjölander. <u>Smarter
Than Man? Intelligence in Whales, Dolphins,
and Humans</u>. Trans. Thomas Teal. New York:
Pantheon, 1972.

Glueck, Nelson. <u>Deities and Dolphins</u>. New York:
Farrar, 1965.

Gormley, Gerard. <u>A Dolphin Summer</u>. New York: Tap-
linger, 1985.

---. Letter to the author. 30 Jan. 1988.

Klinowska, M. "The Cetacean Magnetic Sense—Evidence
from Strandings." <u>Research on Dolphins</u>.
Ed. M. M. Bryden and Richard Harrison. Oxford:
Clarendon, 1986. 401-32.

Lilly, John C. <u>Man and Dolphin</u>. Garden City:
Doubleday, 1961.

---. <u>Programming and Metaprogramming in the Human
Biocomputer</u>. 2nd ed. New York: Julian, 1972.

Linden, Eugene. "Joe and Rosie Go for It." <u>Time</u>
17 Aug. 1987: 72.

McNally, Robert. <u>So Remorseless a Havoc: Of Dolphins,
Whales and Men</u>. Boston: Little, 1981.

The entry for Pyle shows the method for listing a personal interview.

This list of Works Cited includes several essays from the volume *Research on Dolphins*. Generally a writer should not rely upon a single collection. To do so often suggests that the researcher has not ranged widely but has taken a shortcut by obtaining necessary data from one convenient source. It is clear, however, that Curtis Washington has not done that. His research paper refers to many other sources and uses *Research on Dolphins* in several instances only because it contains some of the latest information on pertinent topics.

Washington 23

Morris, Robert J. "The Acoustic Faculty of Dolphins."

Research on Dolphins. Ed. M. M. Bryden and

Richard Harrison. Oxford: Clarendon, 1986.

369–99.

Mowat, Farley. Sea of Slaughter. Boston: Atlantic,

1984.

Pyle, Gregory. Personal interview. 11 Dec. 1987.

Ridgway, S. H. "Dolphin Brain Size." Research on

Dolphins. Ed. M. M. Bryden and Richard

Harrison. Oxford: Clarendon, 1986. 59–70.

"What Is Killing the Atlantic's Dolphins?" Newsweek

24 Aug. 1987: 51.

43j   The APA Style of Documentation

An alternate method of documentation—used widely in the sciences and the social sciences—is that explained in the *Publication Manual of the American Psychological Association* (APA), 1983.

Note that the list of works at the end of the paper is not called "Works Cited" but "References," that in the entries the date of publication comes immediately after the author's name, that initials are used instead of given names, that only the first word of a title is capitalized (except where there is a colon), that no quotation marks are used for titles, and that there are other differences from the method recommended by the Modern Language Association. Follow the preferences of your instructor as to which method you use.

List of references

Include a list entitled "References" immediately after the body of the paper and before the footnotes (if any).

The APA *Manual* specifies that "references cited in text must appear in the reference list; conversely, each entry in the reference list must be cited in text." Alphabetically arranged, the list furnishes publication information necessary for identifying and locating all the references cited in the body of the paper. See the list for the model paper, pp. 447–449, and study the examples below of forms used in APA documentation for various kinds of sources.

A BOOK BY A SINGLE AUTHOR

Fine, L. A. (1990). <u>The souls of the skyscraper:
    Female clerical workers in Chicago, 1870–1930</u>.
    Philadelphia: Temple University Press.

NOTE: Use initials, not first names, of authors. Leave one space between initials. Date of publication in parentheses follows author's name. In the title use capital letters only for the first word, the first word of a subtitle, and any proper nouns. Underline titles of books.

AN EDITED BOOK

Goodale, T. G. (Ed.). (1986). <u>Alcohol and the college student</u>. San Francisco: Jossey–Bass.

A BOOK WITH TWO EDITORS OR AUTHORS

Luloff, A. E., & Swanson, L. E. (Eds.). (1990). <u>American rural communities</u>. Boulder, CO: Westview Press.

NOTE: Authors' last names always come first. An ampersand (&) is used before the name of the last author listed instead of the word *and*. When the name of a state is needed for clarity, use U.S. Postal Service abbreviations.

A BOOK WITH THREE OR MORE EDITORS OR AUTHORS

Albert, E. T., Denise, T. C., & Peterfreund, H. P. (1953). <u>Great traditions in ethics</u>. New York: American Book Company.

A TRANSLATED BOOK

Otto, W. F. (1965). <u>Dionysus: Myth and cult</u>. (R. B. Palmer, Trans.). Bloomington: Indiana University Press.

A MULTIVOLUME WORK

Pederson, L., McDaniel, S. L., Adams, C., & Liao, C. (Eds.). (1989). <u>Linguistic Atlas of the Gulf States</u> (Vol. 3). Athens: University of Georgia Press.

A REPUBLISHED BOOK

Rowson, S. (1986). <u>Charlotte Temple</u>. New York: Oxford University Press. (Original work published 1794)

A REVISED EDITION OF A BOOK

Zabel, M. D. (Ed.). (1951). <u>Literary opinion in America</u> (rev. ed.). New York: Harper.

SECOND OR LATER EDITION OF A BOOK

Trattner, W. I. (1989). <u>From poor law to welfare state: A history of social welfare in America</u> (4th ed.). New York: Free Press.

AN ESSAY IN A COLLECTION OF ESSAYS OR AN ANTHOLOGY

Gongalez, G. A. (1986). Proactive efforts and selected alcohol education programs. In T. G. Goodale (Ed.), <u>Alcohol and the college student</u> (pp. 17–34). San Francisco: Jossey–Bass.

NOTE: Titles of chapters, essays in collections, and articles are not placed in quotation marks nor underlined.

AN INTRODUCTION, FOREWORD, PREFACE, OR AFTERWORD IN A BOOK

Caudill, H. M. (1965). Foreword. In J. E. Weller, <u>Yesterday's people: Life in contemporary Appalachia</u> (pp. xi–xv). Lexington: University Press of Kentucky.

NOTE: The word is *foreword*, not *forward*.

MORE THAN ONE WORK BY THE SAME AUTHOR

Mead, M. (1961). <u>Coming of age in Samoa: A psychological study of primitive youth for Western civilization</u>. New York: Morrow.

Mead, M. (1964). <u>Anthropology, a human science: Selected papers, 1939–1960</u>. Princeton, NJ: Van Nostrand.

NOTE: Works by the same author are arranged according to dates of publication, earlier date first, as above. If works appeared in the same year, arrange alphabetically according to the first letter of the title (disregard *A* or *The*), and add distinguishing letters (*a, b, c,* etc.) to dates in parentheses:

Ornstein, R. E. (1986a). <u>Multimind</u>. Boston: Houghton Mifflin.

Ornstein, R. E. (1986b). <u>The psychology of consciousness</u> (2nd rev. ed.). New York: Penguin Books.

GOVERNMENT PUBLICATION, CORPORATE AUTHOR

United States Department of Labor. (1956). <u>The American workers' fact book</u>. Washington, DC: U.S. Government Printing Office.

ARTICLE IN A MONTHLY PERIODICAL

Edidin, P. (1989, April). Drowning in wealth. <u>Psychology Today</u>, pp. 32–35, 74.

NOTE: Do not abbreviate names of months. Titles of articles are neither underlined nor placed in quotation marks. Discontinuous pages are listed as indicated.

ARTICLE IN A WEEKLY PERIODICAL

Ball, R. (1990, June 11). The new Elizabethans. <u>Time</u>, p. 47.

ARTICLE IN A QUARTERLY PERIODICAL (WITH CONTINUOUS PAGINATION)

Burns, G. (1990). The politics of ideology: The papal struggle with liberalism. <u>American Journal of Sociology, 95</u>, 1123–1152.

NOTE: For this type of entry, the volume number is underlined, and the abbreviation *pp.* is omitted before page numbers.

ARTICLE IN A JOURNAL (WITHOUT CONTINUOUS PAGINATION). TWO AUTHORS

Cannon, H. M., & Morgan, F. W. (1990). A strategic pricing framework. <u>Journal of Services Marketing</u>, <u>4</u>(2), 19–30.

NOTE: When pagination begins anew with each issue, the number of a particular issue must be given. The above entry refers to the second issue in volume 4.

ARTICLE WITH MORE THAN TWO AUTHORS IN A JOURNAL

Fouly, K. A., Bachman, L. F., & Cziko, G. A. (1990). The divisibility of language competence: A confirmatory approach. <u>Language Learning</u>, <u>40</u>, 1–20.

BOOK REVIEW IN A PERIODICAL

Porte, Z. (1990). [Review of <u>Family therapy with</u>
    <u>ethnic minorities</u>]. <u>Journal of Cross-Cultural</u>
    <u>Psychology</u>, <u>21</u>, 250-251.

NOTE: If the review has a title, place it before the material in brackets.

SIGNED ARTICLE IN A DAILY NEWSPAPER

Gordon, L. (1990, May 31). Cutting costs on campus.
    <u>Los Angeles Times</u>, pp. A1, A38, A39.

UNSIGNED ARTICLE (OR EDITORIAL) IN A DAILY NEWSPAPER

Panel urges creation of immigration agency. (1990,
    June 3). <u>Miami Herald</u>, p. 6A.

UNPUBLISHED DISSERTATION

Sweet, S. D. (1989). Liberal education and technology:
    A study of the process of change (Doctoral
    dissertation, University of Denver, 1988).
    <u>Dissertation Abstracts International</u>, <u>49</u>, 3283A.

NOTE: The date after the author's name gives the year of publication of the abstract in *Dissertation Abstracts International.* The second date, 1988, is the year the dissertation was actually completed.

FILM

Wallis, H. B. (Producer), & Huston, J. (Director).
    (1941). <u>The Maltese falcon</u>. [Film]. Warner Bros.

Parenthetical references

As you quote, summarize, or simply refer to sources in the body of your paper, identify these sources on the spot in parentheses. Give the last name of the author (or last names if more than one author) and, after a comma, the date of publication: (Golightly & Walksoftly, 1973). If you are quoting or referring to a specific part of a work, furnish the page number: (Beamish, 1901, p. 34). You can work in this information in various ways, as illustrated below. By using all these formats for parenthetical citations, you will achieve greater flexibility and thus avoid monotony.

> In motion pictures of the silver-screen era, "character development and story were paramount"; in modern films, "those essentials tend to be neglected for the less basic attributes of special effects and 'realistic' language" (Brienza, 1990, p. 183).

> Brienza (1990) observes that in the current cinema important aspects, such as characterization and plot, tend to be neglected. Several other critics (Lamond & Greely, 1983; Secord, 1989; Seltzer, 1990) feel that the art of the motion picture is currently at its highest point.

> According to Brienza (1990), "character development and story were paramount" in motion pictures of the thirties and forties, a contrast to films of our time, in which "those essentials tend to be neglected for the less basic attributes of special effects and 'realistic' language" (p. 183).

For other examples, see the model paper, pp. 429–451. In the APA style, set off (block) quotations of more than forty words. From the left margin indent five spaces; do not use quotation marks; do not indent from the right margin; place the parenthetical reference after the period ending the quotation; do not use a period after the reference. Note the following example:

Many of our struggling young people like to complain about unfairness, about how difficult it is for those without the advantages of inheritances to make it in today's world. They are not the only ones faced with inequalities, however:

> Age has inequalities that are even greater than those of youth. In the matter of income, for example, most of the old are well below the national median, but others are far above it. Many of the great American fortunes are now controlled by aged men (or by their widows or lawyers). (Cowley, 1980, p. 39)

Footnotes

The APA *Manual* advises writers to use footnotes sparingly since all truly pertinent information should be included with parenthetical documentation in the body of the paper. Footnotes are allowed for certain purposes (most of which are not relevant to student writing assignments). You may wish, however, to include "content" footnotes, which add data or information. For an example of such a footnote with commentary, see the model paper, p. 451.

43k Model research paper, APA style

A model research paper in the APA style of documentation with accompanying explanations is given on the following pages.

Beginning with the title page, place a shortened version of your title (or the entire title if it is brief) in the upper right corner of each page. Double-space and type the page number in Arabic numerals. Note that in the APA style, the title page is numbered 1. Use the same heading, and number pages consecutively throughout the paper.

Center the title on the page. APA advises against using in titles "words that serve no useful purposes," such as "A Study of" or "An Experimental Investigation of."

After double-spacing, center your name under the title.

Follow your teacher's instructions in regard to any other information to be included on the title page. Student papers usually include, as here, the name of the course (with a section number if pertinent), the name of the professor, and the date that the paper is submitted. Double-space between all lines.

Spring Break

Alison Cohen

English 101, Professor Cedric Andressen

November 4, 1991

The abstract, usually required when following the APA format, serves as a kind of outline and thesis statement combined. It should not be more than 150 words in length. It should be carefully prepared so that it gives an accurate account of the contents and the purpose of your paper. Do not include in the abstract materials that are not in the text itself. Note: The abstract is a *summary* of the paper, not a commentary on it.

The abstract should be on a separate page. Place the title of the paper (or an abbreviated form of it) in the right-hand corner of the page with the numeral 2 double-spaced below it. Center the word *Abstract*. On the next line, begin at the far left margin. Do not indent.

Abstract

The hundreds of thousands of college students who travel each year to beaches in warm climates during that recess known as spring break have won a reputation for undesirable behavior. Their excessive activities have been widely observed and reported with the result that spring break has come to symbolize to many something new and terrifying: the sudden decay of our youth. A study of ancient civilizations, however, suggests that spring breaks in various forms—often called Saturnalia—have been with us since early times and have been tolerated both because they arise from natural human inclinations and because if limited and controlled they can contribute to the process of maturation.

Double-space between the title (or abbreviated title) in the upper right corner, the page number, the title of the paper (centered), and the first line of text. Leave adequate margins on the right and left as well as at the top and bottom. Indent five spaces for paragraphs.

The opening paragraph effectively describes the phenomenon known as spring break, using materials found in recent accounts. Notice that the author does not try to gloss over the less attractive aspect of the activity. If she is to be convincing, she must convey the impression of objectivity.

The title rather than an author's name is given in the first reference ("Bacchanalia or Bust") because the article is unsigned. Note that in the APA style, the date of publication is an essential part of the parenthetical reference.

The material in the sentence beginning "Wherever they go" is not quoted but paraphrased from Greene. When the author's name is given, the title is not necessary, but the date of publication is included together with the number of the page on which the material appears.

The reference to Greene & Meyer illustrates parenthetical documentation when there is more than one author.

Spring Break

Each year in March or early April hordes of college students put aside their books and worries about classes and grades and set out for beaches usually in the warmer climates of the South or California. Much of the "better heeled college crowd heads for Mexico, the Bahamas, or Key West" ("Bacchanalia or Bust," 1987). Wherever they go, they indulge themselves in various sorts of activities including beer chugs and ketchup-drinking contests (Greene, 1986, p. 30) while enjoying the loud and seemingly endless sounds of rock music. In most colleges and universities this period of revelry is known officially as "spring recess," but to the students it is "spring break," an appropriate designation since there is widespread breakage wherever it occurs. The title of an article in the Chronicle of Higher Education says it all: "In Pursuit of Alcohol and Sex, Thousands Invade Florida and 'Just Go Nuts Down There'" (Greene & Meyer, 1986). The

The sentence beginning "Gerlach (1989)" shows a standard way of quoting in the APA format.

The second paragraph begins with a reference ("such behavior") to the subject discussed in the first paragraph, thus offering an excellent transition to the point to be made in this paragraph: that spring break currently has an extremely bad reputation.

According to APA style, the publication date would not need to be included in connection with Gerlach if this reference had occurred in the same paragraph with the first one. Since the first reference is in the preceding paragraph, the date is given here.

The article in *Newsweek,* from which information is paraphrased here, is not signed. No page reference is furnished because all of the article appears on a single page, which is given in the list of References.

situation at locales outside Florida does not seem to differ substantially. Gerlach (1989) observes that at increasingly popular Padre Island, Texas, "drinking is everywhere, and everything is associated with it" (p. 15). What is "associated with it" is often property damage, especially to hotels and motels. Automobile wrecks and violations are fairly common and arrests widespread.

Such behavior is ample proof to many that we live in an age distinctive for its moral decay. They consider the spring break as evidence that our youth has completely separated itself from time-honored tradition and reached a point of dissolution heretofore unknown. What other conclusions can one reach after reading accounts like those in Gerlach (1989), who comments that during spring break, "student behavior is negative in virtually every possible fashion" (p. 16)? Riordan (1989) describes spring break as "about ten days of debauch on a beach" (p. 16), and an article in Newsweek (1986) reports

The *Publication Manual of the American Psychological Association* (1983) strongly advises against using terms like "the investigator" or "the author," awkward attempts to avoid the first-person pronoun. Such attempts can also be misleading or ambiguous. It is permissible to use "I" as does Alison Cohen here, but use it sparingly. Otherwise, the paper may seem more about yourself than the subject you have investigated.

The paragraph beginning with "I do not wish" constitutes the thesis of the paper. Having described the nature of spring break and commented on its current reputation, Alison Cohen now effectively states what it is she wishes to argue about it. In many papers the thesis statement comes earlier, often in the first paragraph. Given the method of development in this particular paper, it is appropriate in this position.

The final paragraph of this page (continuing onto the next page) begins the segment of the paper offering proof that there is a traditional aspect to spring break, that it is not a new phenomenon but closely resembles celebrations long known in civilized cultures. Alison Cohen's choice of *The Golden Bough* as a source is wise since it is one of the best-known and most reliable works on the subject of cultural patterns over many centuries and in many lands.

Since the author's name, Frazer, is mentioned before the quotation beginning with "annual period," only a page number is needed after it. Contrast this with the next quotation (beginning with "outbursts"), after which both the author's name and the that it is not unusual for fifteen students to occupy one room and that some have lost their lives trying to jump from one hotel or motel balcony to another.

I do not wish to deny the excesses of my fellow students during these vacation periods when they are away from the restraints of home and college, but it is important to correct the idea that spring break is a new phenomenon (and thus represents a dire new trend) and to understand that deplorable as some of the behavior is, the motivation for spring break emerges from needs in human nature that are neither perverse nor alarming.

There is nothing new about a celebration in the midst of busy times in which youthful exuberance is allowed free rein. In fact, such activity has long been known to be part of many ancient civilizations and cultures. In his great work, The Golden Bough (1913), James George Frazer devoted much attention to this "annual period of license, when the customary restraints of law and morality are thrown aside" and

page are given. Note that no publication date is included for Frazer here or in connection with the quotation that follows because the date for *The Golden Bough* has already been given in this paragraph.

Footnote numbers should be slightly raised above the line with no space before them. For an explanation of this footnote, see p. 450.

The paragraph beginning "The basic idea" continues the comparison between spring break and Saturnalia, emphasizing in both a positive aspect, the desire for a sense of community, and a negative one, destructive vitality.

The citation to Hamilton illustrates the basic form of a parenthetical reference, giving the author's last name, date of publication, and the page number for the quotation.

when the celebrants "give themselves up to extravagant mirth and jollity, and when the darker passions find a vent which would never be allowed them in the more staid and sober course of ordinary life" (p. 306). Then, as now, matters usually got out of hand, with "outbursts of the pent-up forces of human nature, too often degenerating into wild orgies of lust and crime" (Frazer, p. 306). Just as spring break takes place toward the end of the school year, so did these traditional revelries occur during the end of the calendar year. "Now, of all these periods of license," writes Frazer, "the one which is best known and which in modern languages has given its name to the rest, is the Saturnalia" (p. 306).[1]

The basic idea behind the Saturnalia was one of fellowship and goodwill, of not only enjoying oneself but also others' company. It was declared a time of peace rather than conflict, and it stressed "the idea of equality . . . when all were on the same level" (Hamilton, 1940, p. 45). No attitude seems more

The paragraph beginning with "Yet it has seemed" illustrates how a saturnalian experience can be important in the process of maturation by discussing one of Shakespeare's heroes, Prince Hal, later Henry V, King of England.

Note that the names of plays are underlined.

pervasive among the spring break crowds today than this very spirit of camaraderie and equality. But, as C. L. Barber (1951) points out, "a saturnalian attitude, assumed by a clear-cut gesture toward liberty, brings with it an accession of 'wanton' vitality. . . . The energy normally occupied in maintaining inhibition is freed for celebration" (p. 598). "Wanton vitality" was the main cause of Saturnalia's getting out of hand in ancient times as it is the reason behind such disorder and misbehavior among spring breakers today.

Yet it has seemed important since early times to have Saturnalia. Why? The answer to this question is that such celebrations, subject as they are to becoming riotous and immoral, can nevertheless be basically healthy because they can serve a positive function in the total makeup of human beings. In one of his plays, Henry IV, Part I, Shakespeare depicts his hero, Prince Hal, as enjoying one long spring break as he rollicks in the company of disreputable

In this paper two works by the same author, C. L. Barber, are cited. No confusion results, however, because the date of publication given here, 1955, distinguishes this work from the one cited in the previous paragraph with the date 1951.

The concluding paragraph is not an unqualified defense of spring break since it suggests the need for restraint, limitations on freedom. Nevertheless, it forcefully makes the final points that the need for such celebrations appears to emerge from healthy sources within, that this motivation has long been recognized, and that spring break in some form will probably always exist.

The quotation "limit holiday" repeats Barber's words in the preceding paragraph, which was documented, and therefore needs no citation here. However, Alison Cohen has chosen to add emphasis by underlining the word *limit*. The words *italics mine* are added in parentheses to indicate that the original quotation has been altered in this way.

The two sentences beginning with "In defining Saturnalia" illustrate two ways of citing sources, the first giving the name of the author before the quotation, the second after it in parentheses.

Spring Break

8

characters. When the time comes for him truly to grow up, however, and to assume power as King of England, he becomes one of that country's great rulers—better for his long spring break with Falstaff and his band of merrymakers. Shakespeare "dramatizes not only holiday but also the need for holiday and the need to limit holiday" (Barber, 1955, p. 28). A time of "release," according to Barber, leads to a time of "clarification" (p. 25).

Spring break may not be instrumental in forming many kings or presidents these days, and there certainly is no excuse for the total lack of responsibility and discipline that some students exhibit during these periods. Many fail to "limit holiday" (italics mine), but for those who do, spring break may not be all bad either. In defining Saturnalia, J. E. Cirlot (1971) has written that it embodies "a desire to concentrate into a given period of time all the possibilities of existence" (p. 279). It is a quest for a "way out of time" (Cirlot, p.

279). Today's spring breakers probably do not realize that they have these deep-seated motivations, but they do know that there is something about this experience like none other. The desire to participate in it has been with humankind since early civilizations and is likely to be with us for a long time to come.

Begin on a new page for the list of sources called, in the APA style, *References*. Center the word on the page, double-space, and list entries in alphabetical order. Each new entry begins at the left margin; indent additional lines in individual entries three spaces.

Since the entry for "Bacchanalia or bust" lists an unsigned article, the title is given first (and the entry alphabetized in the list according to the first letters of the title). This entry shows the APA style for indicating articles in weekly magazines.

The two works by the same author, C. L. Barber, are listed according to the date of publication, the one with the earlier date first. The first entry for Barber illustrates how to list an article in a journal with continuous pagination throughout a volume. When listing a title, capitalize only the first letter of the first word, of a word coming after a colon, and of any proper noun. Do not place titles of articles in quotation marks. Underline titles of books. The second entry for Barber illustrates the form for an article or chapter in a book with an editor.

The entry for Cirlot shows how to indicate a translated work.

Alison Cohen used one volume of the several that make up *The Golden Bough* and an edition later than the first. Note the proper way of presenting this information.

References

Bacchanalia or bust. (1987, April 18). The Economist,

p. 32.

Barber, C. L. (1951, Autumn). The saturnalian pattern

in Shakespeare's comedy. Sewanee Review, 59,

593–611.

Barber, C. L. (1955). From ritual to comedy: An

examination of Henry IV. In W. K. Wimsat, Jr.

(Ed.), English stage comedy (pp. 22–51). New York:

Columbia University Press.

Beers, busts, and tragedy. (1986, April 7). Newsweek,

p. 70.

Cirlot, J. E. (1971). A dictionary of symbols (J.

Sage, Trans.). New York: Philosophical Library.

Frazer, J. G. (1913). The golden bough: A study in

magic and religion (3rd ed.). (Vol. 9). London:

Macmillan.

Gerlach, J. (1989, Spring). Spring break at Padre

Island: A new kind of tourism. Focus, pp. 13–16,

29.

The entry for Greene & Meyer shows the form for a work with two authors.

Spring Break

11

Greene, E. (1986, April 2). At the Button in Fort
   Lauderdale. Chronicle of Higher Education, pp. 1,
   30.

Greene, E., & Meyer, T. J. (1986, April 2). In pursuit
   of alcohol and sex, thousands invade Florida and
   "just go nuts down there." Chronicle of Higher
   Education, pp. 29–30.

Hamilton, E. (1940). Mythology. New York: Mentor
   Books.

Riordan, T. (1989, March 27). Miller guy life. New
   Republic, pp. 16–17.

Begin a new page for footnotes, and place it last in your paper after the list of References. Be sure you have ordered your pages in accordance with the following checklist:

1. title page
2. abstract
3. body of text
4. references
5. footnotes

The footnote here is what APA refers to as a "content footnote," one that serves to "supplement or amplify substantive information in the text" (p. 105). Though not closely enough related to the line of argument in the paper itself to warrant inclusion there, the material in this note nevertheless contributes to the interest and background of the argument. It is not always necessary to have footnotes. The APA *Manual* cautions against the inclusion of notes with "irrelevant or nonessential information" (p. 105).

Indent five spaces for each new footnote. Footnotes are numbered consecutively throughout with numerals raised slightly above the line and without periods or spaces after them.

Footnote

[1]Writers who report on the activities of spring break often use terms like <u>rites of spring</u> and <u>ritual</u>. Greene and Meyer (1986) point out that some aspects of spring break resemble the forms of a religion (p. 31). It is not uncommon to encounter the term <u>bacchanalia</u> or <u>bacchanalian</u> in accounts of spring breaks. <u>Saturnalia</u> is a more inclusive term than <u>bacchanalia</u> and incorporates more positive elements of the experience.

Glossary of
Usage

44 Glossary of Usage *gl / us*

Many items not listed here are covered in other sections of this book and may be located through the index. For words found neither in this glossary nor in the index, consult a good dictionary.

A, an   Use *a* as an article before consonant sounds; use *an* before vowel sounds.

a nickname

a house [the *h* is sounded]

a historical novel [though the British say *an*]

a union [long *u* has the consonant sound of *y*]

an office

an hour [the *h* is not sounded]

an honor

an uncle

Accept, except   As a verb, *accept* means "to receive"; *except* means "to exclude." *Except* as a preposition also means "but."

Every legislator *except* Mr. Whelling refused to *accept* the bribe.

We will *except* (exclude) this novel from the list of those to be read.

Accidently   A misspelling usually caused by mispronunciation. Use *accidentally*.

Advice, advise   Use *advice* as a noun, *advise* as a verb.

Affect, effect   *Affect* is a verb meaning "to act upon" or "to influence." *Effect* may be a verb or a noun. *Effect* as a verb means "to cause" or "to bring about"; *effect* as a noun means "a result," "a consequence."

The patent medicine did not *affect* (influence) the disease.

The operation did not *effect* (bring about) an improvement in the patient's health.

The drug had a drastic *effect* (consequence) on the speed of the patient's reactions.

Aggravate   Informal in the sense of "annoy," "irritate," or "pester." Formally, it means to "make worse or more severe."

Agree to, agree with   *Agree to* a thing (plan, proposal); *agree with* a person

He *agreed to* the insertion of the plank in the platform of the party.

He *agreed with* the senator that the plank would not gain many votes.

Ain't   Nonstandard or slang for *is not* or *are not.*

All ready, already   *All ready* means "prepared, in a state of readiness"; *already* means "before some specified time" or "previously" and describes an action that is completed.

> The riders were *all ready* to mount. (fully prepared)
>
> Mr. Bowman had *already* bagged his limit of quail. [action completed at time of statement]

All together, altogether   *All together* describes a group as acting or existing collectively; *altogether* means "wholly, entirely."

> The sprinters managed to start *all together.*
>
> I do not *altogether* approve of the decision.

Allusion, illusion   An *allusion* is a casual reference. An *illusion* is a false or misleading sight or impression.

Alot   Nonstandard for *a lot.*

Alright   Nonstandard for *all right.*

Among, between   *Among* is used with three or more persons or things; *between* is used with only two.

> It will be hard to choose *among* so many candidates.
>
> It will be hard to choose *between* the two candidates.

Amongst   Pretentious. Prefer *among.*

Amount, number   *Amount* refers to mass or quantity; *number* refers to things that may be counted.

> That is a large *number* of turtles for a pond that has such a small *amount* of water.

An   See **A.**

And etc.   See **Etc.**

And/or   A shortcut found chiefly in legal documents. Avoid.

Anxious, eager   *Anxious* is not a synonym for *eager. Anxious* means "worried or distressed"; *eager* means "intensely desirous."

> The defendant was *anxious* about the outcome of the trial.
>
> Most people are *eager* to hear good news.

Anyways   Prefer *anyway.*

Anywheres   Prefer *anywhere.*

As   Weak or confusing in the sense of *because*.

>   The client collected the full amount of insurance *as* her car ran off
>   the cliff and was totally demolished.

At this (*or* that) point in time   Avoid. Wordy and trite.

Awesome   Informal for *impressive*.

Awful   Informal for *bad, shocking, ludicrous, ugly*.

Awhile, a while   *Awhile* is an adverb; *a while* is an article and a noun.

>   Stay *awhile*.

>   Wait here for *a while*. [object of preposition]

Bad, badly   See p. 77.

Basic and fundamental   Redundant. Use one or the other.

Because   See **Reason is because**.

Being as, being that   Use *because* or *since*.

Beside, besides   *Beside* means "by the side of," "next to"; *besides* means
"in addition to."

>   Mr. Potts was sitting *beside* the stove.

>   No one was in the room *besides* Mr. Potts.

Between   See **Among**.

Between you and I   Wrong case. Use *between you and me*.

Bottom line   Informal for "the basic or most important factor or consideration." Avoid in formal writing.

Bring, take   Use *bring* to indicate movement toward the place of speaking
or regarding. Use *take* for movement away from such a place.

>   *Take* this coupon to the store and *bring* back a free coffee mug.

Bug   Informal or slang in almost every sense except when used to name
an insect.

Bunch   Informal for a group of people.

Bust, busted   Slang as forms of *burst*. *Bursted* is also unacceptable.

Can, may   In formal English, *can* is still used to denote ability; *may*, to
denote permission. Informally the two are interchangeable.

>   FORMAL
>   *May* [not *can*] I go?

Capital, capitol   *Capitol* designates "a building that is a seat of government"; *capital* is used for all other meanings.

Center around   Use *center in* (or *on*) or *cluster around*.

Climactic, climatic   *Climactic* pertains to a climax; *climatic* pertains to climate.

Compare, contrast   Do not use interchangeably. *Compare* means to look for or reveal likenesses; *contrast* treats differences.

Compare to, compare with   In formal writing use *compare to* when referring to similarities and *compare with* when referring to both similarities and differences.

> The poet *compared* a woman's beauty *to* a summer's day.

> The sociologist's report *compares* the career aspirations of male undergraduates *with* those of female undergraduates.

Complement, compliment   As a verb, *complement* means to "complete" or "go well with"; as a noun, it means "something that completes." *Compliment* as a verb means "to praise"; as a noun it means "an expression of praise."

> Her delicate jewelry was an ideal *complement* to her simple but tasteful gown.

> Departing guests should *compliment* a gracious hostess.

Conscience, conscious   *Conscience,* a noun, refers to the faculty that distinguishes right from wrong. *Conscious,* an adjective, means "being physically or psychologically aware of conditions or situations."

> His *conscience* dictated that he return the money.

> *Conscious* of their hostility toward him, the senator left the meeting early.

Consensus of opinion   Wordy. Prefer *consensus* or *opinion*.

Continual, continuous   Use *continual* to refer to actions that recur at intervals, *continuous* to mean "without interruption."

> The return of the bats to the lake every evening is *continual*.

> Breathing is a *continuous* function of the body.

Contractions   Avoid contractions *(don't, he's, they're)* in formal writing.

Contrast   See **Compare**.

Cool   Slang when used to mean "excellent" or "first-rate."

Could care less   Nonstandard for "could not care less."

Could of, would of   See **Of.**

Couple of   Informal for "two" or "a few."

Criteria, phenomena, data   Plurals. Use *criterion, phenomenon* for the singular. *Data,* however, can be considered singular or plural.

Cute, great, lovely, wonderful   Avoid as substitutes for more exact words of approval.

Different from, different than   Prefer *different from* in your writing.

> Children's taste in music is often much *different from* that of their parents.

Differ from, differ with   *Differ from* means "to be unlike"; *differ with* means "to disagree."

> The economic system of Poland *differs from* that of New Zealand.

> Poland *differs with* some neighboring countries on economic policy.

Discreet, discrete   *Discreet* means "tactful"; *discrete* means "separate" or "distinct."

Disinterested, uninterested   Use *disinterested* to mean someone is "unbiased, objective"; use *uninterested* to mean someone is "not interested, indifferent."

> A *disinterested* person is helpful in solving disputes.

> *Uninterested* members of an audience seldom remember what a speaker says.

Don't   Contraction of *do not;* not to be used for *doesn't,* the contraction of *does not.*

Double negative   Avoid such nonstandard phrases as *don't have no, can't hardly,* and so on.

Due to   Objectionable to some when used as a prepositional phrase meaning *because of.*

OBJECTIONABLE
Due to the laughter, the speaker could not continue.

BETTER
Because of the laughter, the speaker could not continue.

Each and every   Redundant. Use one or the other, not both.

Each other, one another   When two persons are involved, use the expression *each other;* when three or more persons are involved, use the expression *one another.*

The poet and her editor wrote *each other* frequently.

Arguing heatedly with *one another,* the members of the jury deliberated the case.

Eager See **Anxious.**

Early on Redundant. Use *early.* Omit *on.*

Effect See **Affect.**

Elicit, illicit *Elicit,* a verb, means "to evoke or bring forth." *Illicit,* an adjective, means "not permitted" or "unlawful."

A pianist's innovative style will sometimes *elicit* a negative response from the critics.

Authorities discovered his participation in an *illicit* trade involving weapons.

Emigrate, immigrate *To emigrate* means "to leave one's country and settle in another." *To immigrate* means "to enter another country and reside there."

Luis *emigrated from* Spain to the United States when he was only five years old.

My grandmother *immigrated to* this country from the city of Messina, Sicily, in 1919.

Eminent, imminent *Eminent* means "distinguished"; *imminent* means "about to happen."

Enthused Use *enthusiastic* in formal writing.

Equally as good Redundant. Use *equal to* or *as good as.*

Etc. Do not use *and etc. Etc.* means "and so forth" or "and others."

Ever, every Use *every* in *everybody, every day, every now and then, every other;* use *ever* in *ever and anon, ever so humble.*

Exam Considered informal by some authorities. *Examination* is always correct.

Except See **Accept.**

Expect Informal for *believe, suppose, suspect, think,* and so forth.

Explicit, implicit *Explicit* means "directly expressed or clearly defined." *Implicit* means "implied or understood."

Hitler confronted Neville Chamberlain with *explicit* demands.

The Prime Minister did not have to answer; his disappointment was *implicit.*

Fabulous   Informal for *extremely pleasing.*

Fantastic   Informal for *extraordinarily good.*

Farther, further   Generally interchangeable, though many persons prefer *farther* in expressions of physical distance and *further* in expressions of time, quantity, and degree.

> My car used less gasoline and went *farther* than his.

> The second speaker went *further* into the issues than the first.

Feel like   Nonstandard for "feel that."

> NOT
> I feel like I could have been elected.

> CORRECT
> I feel that I could have been elected.

See **Like.**

Fewer, less   Use *fewer* to denote number; *less,* to denote amount or degree.

> With *fewer* advertisers, there will be *less* income from advertising.

Finalize   Bureaucratic. Avoid.

Fine   Avoid as a substitute for a more exact word of approval or commendation.

Fix   Informal for the noun *predicament.*

Flunk   Informal: prefer *fail* or *failure* in formal usage.

Fun   Informal as an adjective.

> NOT
> Go to Acapulco for a *fun* vacation.

Funny   Informal for *strange, remarkable,* or *peculiar.*

Further   See **Farther.**

Good   Incorrect as an adverb. See **10.**

Got to   In formal writing prefer *have to, has to,* or *must.*

> People must [not *got to*] understand that voting is a great privilege.

Great   Informal for "first-rate."

Hanged, hung   Both *hanged* and *hung* are the past tense and past participle forms of *hang,* but they are not interchangeable. *Hanged* means "executed by hanging"; *hung* means "suspended."

In modern times criminals are seldom *hanged.*

The decorator *hung* a reproduction of Picasso's *Portrait of a Woman* in the living room.

Hardly   See **Not hardly.**

He, she   Traditionally, *he* has been used to mean *he* or *she*. Today this usage is unacceptable. For additional discussion, alternatives, and examples, see pp. 60–61.

He/she, his/her   Shortcuts often used in legal writing. For more acceptable ways to avoid sexism in use of pronouns, see pp. 60–61.

Himself   See **Myself.**

Hopefully   When *hopefully* is used as a sentence modifier in the sense "it is hoped," it is often unclear who is doing the hoping—the writer or the subject.

VAGUE
Hopefully, the amendment will be approved.

CLEAR
Most local voters hope that the amendment will be approved.

CLEAR (but with a different meaning)
I hope the amendment will be approved.

If, whether   Use *if* to indicate a condition; use *whether* to specify alternatives.

The reception will be held outdoors *if* it does not rain.

The reception will be held indoors *whether* it rains or not.

Illusion   See **Allusion.**

Impact on   Sociological and bureaucratic jargon. Avoid *impact* as a verb.

Imply, infer   *Imply* means "to hint" or "suggest"; *infer* means "to draw a conclusion."

The speaker *implied* that Mr. Dixon was guilty.

The audience *inferred* that Mr. Dixon was guilty.

In, into   *Into* denotes motion from the outside to the inside; *in* denotes position (enclosure).

The lion was *in* the cage when the trainer walked *into* the tent.

Individual   Avoid using for *person.*

Infer   See **Imply.**

Input Avoid in formal writing as a substitute for "information" or "opinion."

In regards to Unidiomatic: use *in regard to* or *with regard to*.

Interact with Overused and vague. Avoid for more precise language.

Interface with Sociological and bureaucratic jargon. Avoid.

Into See **In.**

Irregardless Nonstandard for *regardless*.

Is when, is where Ungrammatical use of adverbial clause after a linking verb. Often misused in definitions and explanations.

NONSTANDARD

Combustion *is when* [or *is where*] oxygen unites with other elements.

STANDARD

Combustion occurs when oxygen unites with other elements.

Combustion is a union of oxygen with other elements.

Its, it's *Its* is the possessive case of the pronoun *it; it's* is a contraction of *it is* or *it has*.

It's exciting to parents when their children succeed.

The dog eagerly chased *its* tail.

Kind of, sort of Informal as adverbs: use *rather, somewhat,* and so forth.

INFORMAL

Mr. Josephson was *sort of* disgusted.

FORMAL

Mr. Josephson was *rather* disgusted.

FORMAL (not an adverb)

What *sort of* book is that?

Kind of a, sort of a Delete the *a;* use *kind of* and *sort of*.

What *kind of* [not *kind of a*] pipe do you smoke?

Lay, lie See p. 40.

Lead, led *Lead* is an incorrect form for the past tense *led*.

Learn, teach *Learn* means "to acquire knowledge." *Teach* means "to impart knowledge."

She could not *learn* how to work the problem until Mrs. Smithers *taught* her the formula.

Less See **Fewer.**

Liable See **Likely.**

Lie See p. 40.

Like A preposition. Instead of *like* as a conjunction, prefer *as, as if,* or *as though.*

PREPOSITION
> She acted *like* a novice.

CONJUNCTION
> She acted *as if* she had never been on the stage before. [correct]

> She acted *like* she had never had a date before. [informal]

Such popular expressions as "tell it like it is" derive part of their appeal from their lighthearted defiance of convention.
Do not use *like* [the verb] for *lack.*
Do not use *like* for *that* as in *feel like.*
See **Feel like.**

Likely, liable Use *likely* to express probability; use *liable,* which may have legal connotations, to express responsibility or obligation.

> You are *likely* to have an accident if you drive recklessly.

> Since your father owns the car, he will be *liable* for damages.

Literally Do not use for *figuratively.*

> She was *literally* walking on clouds. [Means that she was *actually* stepping from cloud to cloud.]

Loose, lose Frequently confused. *Loose* is an adjective; *lose* is a verb.

> She wore a *loose* and trailing gown.

> Speculators often *lose* their money.

Lot of, lots of Informal in the sense of *much, many, a great deal.*

Lovely See **Cute.**

Mad Informal when used to mean "angry."

Man (mankind) Objectionable to many. Prefer *humankind* or *human beings.*

May See **Can.**

Media Plural of *medium.* Do not use to refer to a single entity, such as the press or television.

Moral, morale *Moral* refers to a lesson; *morale,* to mood or spirit.

Most   Informal for *almost* in such expressions as the following:

He is late for class *almost* [not *most*] every day.

Myself, yourself, himself, herself, itself   These words are reflexives or intensives, not strict equivalents of *I, me, you, he, she, him, her, it.*

INTENSIVE

I *myself* helped Father cut the wheat.

I helped Father cut the wheat *myself.*

REFLEXIVE

I cut *myself.*

NOT

The elopement was known only to Sherry and *myself.*

BUT

The elopement was known only to Sherry and *me.*

NOT

Only Kay and *myself* had access to the safe.

BUT

Only Kay and *I* had access to the safe.

Neat   Overused and informal in the sense of "pleasing" or "appealing."

NOT

It would be *neat* to be able to predict the future.

Nice   A weak substitute for more exact words like *attractive, modest, pleasant, kind,* and so forth.

Not hardly   Double negative. Avoid.

Nowhere near   Informal for "not nearly."

Mount Whitney is *nowhere near* [use *not nearly*] as tall as Everest.

Nowheres   Dialectal for *nowhere.*

Nucular   A misspelling and mispronunciation of *nuclear.*

Number   See **Amount.**

Of   Not to be used for "have."

Had he lived, the emperor could *have* [not *of*] prevented the war.

The physician would *have* [not *of*] come had he known of the illness.

Off of   *Off* is sufficient.

He fell *off* [not *off of*] the water tower.

O.K., OK, okay   Informal.

Per   Do not use for "a" or "an" in formal writing.

> Guests paid fifty dollars *a* [not *per*] plate at the political dinner.

Percent, percentage   *Percent* refers to a specific number: 8 percent. *Percentage* is used when no number is specified: a *percentage* of the stock. The % sign after a percentage (8%) is acceptable in business and technical writing. Write out the word *percent* in formal writing. The number before percent is given in figures except at the start of a sentence.

> Employees own *8 percent* of the stock.

> *Eight percent* of the stock is owned by employees.

Percent can be singular or plural. See **7g.**
A *percentage* takes a singular or a plural verb. See **7g.**

Phenomena   See **Criteria.**

Photo   Informal.

Plus   Avoid using for the conjunction *and.*

Precede, proceed   *Precede* means "to come before." *Proceed* means "to move onward."

> Freshmen often *precede* seniors on the waiting list for campus housing.

> The instructor answered a few questions and then *proceeded* with the lecture.

Principal, principle   Use *principal* to mean "the chief" or "most important." Use *principle* to mean "a rule" or "a truth."

> The *principal* reason for her delinquency was never discussed.

> The *principal* of Brookwood High School applauded.

> To act without *principle* leads to delinquency.

Provided, providing   *Provided,* the past participle of the verb *provide,* is used as a conjunction to mean "on condition that." *Providing,* the present participle of the verb *provide,* is often mistakenly used for *provided.*

> The cellist will compete in the Canadian Music Festival *provided* (not *providing*) she wins the regional competition.

Quote   A verb; prefer *quotation* as a noun.

Raise, rise   See p. 40.

Real   Informal or dialectal as an adverb meaning *really* or *very*.

Reason is (was) because   Use *the reason is (was) that*. Formally, *because* should introduce an adverbial clause, not a noun clause used as a predictate nominative.

NOT

The *reason* Abernathy enlisted *was because* he failed in college.

BUT

The *reason* Abernathy enlisted *was that* he failed in college.

OR

Abernathy enlisted *because* he failed in college.

Relate to   Overused and imprecise in the sense of "to understand," "to appreciate," or "to sympathize with."

NOT

Candidates promising higher taxes are not popular; I can relate to that.

Respectfully, respectively   *Respectfully* means "with respect"; *respectively* means "each in the order given."

He *respectfully* thanked the president for his diploma.

Crossing the platform, he passed *respectively* by the speaker, the dean, and the registrar.

Revelant   A misspelling and mispronunciation of *relevant*.

Sensual, sensuous   *Sensual* connotes the gratification of bodily pleasures; *sensuous* refers favorably to what is experienced through the senses.

Set, sit   See p. 40.

Shall, will   In strictly formal English, to indicate simple futurity, *shall* is conventional in the first person (I *shall*, we *shall*); *will*, in the second and third persons (you *will*, he *will*, they *will*). To indicate determination, duty, or necessity, *will* is formal in the first person (I *will*, we *will*); *shall*, in the second and third persons (you *shall*, he *shall*, they *shall*). These distinctions are weaker than they used to be, and *will* is increasingly used in all persons.

Should of   See **Of**.

So, so that   Use *so that* instead of *so* to express intent or purpose.

Most people use credit cards *so that* (not *so*) they can pay their bills at the end of the month.

Sometime, some time   *Sometime* is used adverbially to designate an indefinite point of time. *Some time* refers to a period or duration of time.

I will see you *sometime* next week.

I have not seen him for *some time.*

Sort of   See **Kind of.**

Sort of a   See **Kind of a.**

Specially   Do not use for *especially.*

Iced tea is *especially* [not *specially*] welcome on a hot day.

Stationary, stationery   An adjective, *stationary* means "not moving." A noun, *stationery* refers to writing materials or office supplies.

Super   Informal for *excellent.*

Sure   Informal as an adverb for *surely, certainly.*

INFORMAL
The speaker *sure* criticized his opponent.

FORMAL
The speaker *certainly* criticized his opponent.

Sure and, try and   Use *sure to, try to.*

Be *sure to* (not *sure and*) notice the costumes of the Hungarian folk dancers.

Suspicion   Avoid as a verb; use *suspect.*

Teach   See **Learn.**

Than, then   Do not use one of these words for the other. *Than,* a conjunction, is used in comparisons. *Then,* an adverb, relates to time.

Calculus is more complicated *than* algebra.

Drizzle the pizza with oil; *then* bake it at 350°.

That, which   When introducing a relative clause essential to the meaning of the sentence (a restrictive clause), use *that.* Use *which* in a relative clause that provides the sentence with additional, but not necessary, information (a nonrestrictive clause). Note that the nonrestrictive clause is set off with commas.

The driver of the car *that* ran the red light has been fined.

The jacket, *which* is several years old, has large pockets and a detachable hood.

Their, there Not interchangeable; *their* is the possessive of *they; there* is either an adverb meaning "in that place" or an expletive.

> *Their* dachshund is sick.

> *There* it is on the corner. [adverb of place]

> *There* is a veterinarian's office in this block. [expletive]

These (those) kind, these (those) sort *These (those)* is plural; *kind (sort)* is singular. Therefore use *this (that) kind, this (that) sort; these (those) kinds, these (those) sorts.*

Thusly Prefer *thus.*

Toward, towards Although both are acceptable, prefer *toward.*

Try and See **Sure and.**

Uninterested See **Disinterested.**

Unique Means "one of a kind"; hence it may not logically be compared. *Unique* should not be loosely used for *unusual* or *strange.*

Use Sometimes carelessly written for the past tense, *used.*

> Thomas Jefferson *used* [not *use*] to bathe in cold water almost every morning.

Utilize Bureaucratic substitute for the shorter, and more acceptable, *use.*

Wait on Unidiomatic for *wait for. Wait on* correctly means "to serve."

Ways Prefer *way* when designating a distance.

> NOT
> a long *ways*

> BUT
> a long *way*

When, where See **Is when.**

Where Do not misuse for *that.*

> I read in the newspaper *that* [not *where*] you saved a child's life.

While Do not use *while* in a sentence when it can be taken to mean "although."

> AMBIGUOUS
> *While* I prepared dinner, Ann did nothing.

Who, which *Who* refers to people; *which* refers to things.

Some listeners are amused at a newscaster *who* cannot pronounce names correctly.

Copies of the author's last novel, *which* was published in 1917, are now rare.

Whose, who's   *Whose* is the possessive of *who; who's* is a contraction of *who is*.

Wicked   Slang when used for *excellent* or *masterly*.

-wise   A suffix overused in combinations with nouns, such as *budgetwise, businesswise,* and *progresswise*.

Wonderful   See **Cute**.

Would of   See **Of**.

You know   Annoying, repetitious, and meaningless when used as a question inserted in a sentence.

Avoid: He was just scared, *you know?*

Glossary of
Grammatical
Terms

45 Glossary of Grammatical Terms *gl / gr*

This is by no means a complete list of terms used in discussing grammar. See "Grammar" (pp. 2–26), the index, other sections of this book, and dictionaries.

Absolute phrase   See **20L.**

Active voice   See **Voice.**

Adjectival   A term describing a word or word group that modifies a noun.

Adjective   A word that modifies a noun or a pronoun. See pp. 8–9.

> *Her young* horse jumped over *that high* barrier for *the first* time.

Adjective clause   See **Dependent clause.**

Adverb   A word that modifies a verb, an adjective, or another adverb (see pp. 9–10).

Adverbial clause   See **Dependent clause.**

Agreement   The correspondence between words in number, gender, person, or case. A verb agrees in number and person with its subject. A pronoun must agree in number, person, and gender with its antecedent.

Antecedent   A word to which a pronoun refers.

> *antecedent* *pronoun*
> ↓ ↓
> When the ballet *dancers* appeared, *they* were dressed in pink.

Appositive   A word, phrase, or clause used as a noun and placed beside another word to explain it.

> *appositive*
> ↓
> The poet *John Milton* wrote *Paradise Lost* while he was blind.

Article   *A* and *an* are indefinite articles; *the* is the definite article.

Auxiliary verb   A verb used to help another verb indicate tense, mood, or voice. Principal auxiliaries are forms of the verbs *to be* and *to do*. See p. 7.

> I *am* studying.
> I *do* study.

I *shall* go there next week.
He *may* lose his job.

Case   English has remnants of three cases: subjective, possessive, and objective. Nouns are inflected for case only in the possessive *(father, father's)*. An alternative way to show possession is with the "of phrase" *(of the house)*. Some pronouns, notably the personal pronouns and the relative pronoun *who,* are still fully inflected for three cases:

SUBJECTIVE (acting)
    I, he, she, we, they, who

POSSESSIVE (possessing)
    my (mine), your (yours), his, her (hers), its, our (ours), their (theirs), whose

OBJECTIVE (acted upon)
    me, him, her, us, them, whom

Clause   A group of words containing a subject and a predicate. See **Independent clause; Dependent clause.** See pp. 24–25.

Collective noun   A word identifying a class or a group of persons or things. See p. 3.

Comma splice (or comma fault)   An error that occurs when two independent clauses are incorrectly linked by a comma with no coordinating conjunction. See pp. 32–34.

Comparative and superlative degrees   See **10a** and **10b.**

Complement   A word or group of words used to complete a predicate. Predicate adjectives, predicate nominatives, direct objects, and indirect objects are complements. See pp. 19–20.

Complex, compound, compound-complex sentences   A *complex sentence* has one independent clause and at least one dependent clause. A *compound sentence* has at least two independent clauses. A *compound-complex sentence* has two or more independent clauses and one dependent clause or more. See p. 26.

Compound subjects   See p. 16.

Conjugation   The inflection of the forms of a verb according to person, number, tense, voice, and mood. See the abbreviated form of the conjugation of the verb *walk* in **4.**

Conjunction   A word used to connect sentences or sentence parts. See also **Coordinating conjunction, Correlative conjunctions, Subordinating conjunction,** and p. 11.

Conjunctive adverb An adverb used to relate two independent clauses that are separated by a semicolon: *besides, consequently, however, moreover, then, therefore,* and so on. (See **22a**.)

Contraction The shortening of two words combined by replacing omitted letters with an apostrophe.

>*I've* for *I have. Isn't* for *is not.*

Coordinate clause See **Independent clause.** When there are two independent clauses in a compound or a compound-complex sentence, they may be called coordinate clauses.

Coordinating conjunction A simple conjunction that joins sentences or parts of sentences of equal rank *(and, but, for, nor, or, so, yet).* See p. 11.

Correlative conjunctions Conjunctions used in pairs to join coordinate sentence elements. The most common are *either . . . or, neither . . . nor, not only . . . but also, both . . . and.*

Dangling modifier A modifier that is not clearly attached to a word or element in the sentence.

DANGLING MODIFIER
>Following a regimen of proper diet and exercise, Alan's weight can be controlled.

REVISED
>Following a regimen of proper diet and exercise, Alan can control his weight.

Declension The inflection of nouns, pronouns, and adjectives in case, number, and gender. See especially **9**.

Degrees (of modifiers) See **10a** and **10b**.

Demonstrative adjective or pronoun A word used to point out *(this, that, these, those).*

Dependent (subordinate) clause A group of words that contains both a subject and a predicate but that does not stand alone as a sentence. A dependent clause is frequently signaled by a subordinator *(because, since, that, what, who, which,* and so on) and always functions as an adjective, adverb, or noun.

>*adjective clause*
>The tenor *who sang the aria* had just arrived from Italy.

>*noun clause*
>The critics agreed *that the young tenor had a magnificent voice.*

adverb clause
When he sang, even the sophisticated audience was enraptured.

Diagramming   Diagramming uses systems of lines and positioning of words to show the parts of a sentence and the relationships between them. Its purpose is to make understandable the way writing is put together. (See the example below.)

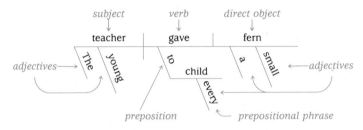

Direct object   A noun, pronoun, or other substantive that receives the action of the verb. See p. 19.

The angler finally caught the old *trout.*

Double negative   Nonstandard use of two negative words within the same sentence.

DOUBLE NEGATIVE
I do *not* have *hardly* any problems with my car.

REVISED
I have *hardly* any problems with my car.

Elliptical clause   A clause in which one or more words are omitted but understood.

understood

The director admired no one else as much as (*he admired* or *he did*) Faith DeFelce.

Expletive   See **7h.**

Faulty predication   Errors occur in predication when a subject and its complement are put together in such a fashion that the sentence is illogical or unmeaningful. See **15b.**

FAULTY
A reporter taking pictures meant that there would be publicity. [It is illogical to say that *taking pictures meant.*]

BETTER

> Since a reporter was taking pictures, we knew that there would be publicity.

Fragment   Part of a sentence written and punctuated as a complete sentence.

FRAGMENT

> *Jostling and bantering with one another.* The team headed for the locker room.

REVISED

> Jostling and bantering with one another, the team headed for the locker room.

Fused (or run-on) sentence   An error that occurs when two independent clauses have neither punctuation nor coordinating conjunctions between them.

FUSED

> The average adult has about twelve pints of blood this amount is roughly 9 percent of total body weight.

REVISED

> The average adult has about twelve pints of blood; this amount is roughly 9 percent of total body weight.

Gender   The classification of nouns and pronouns into masculine, feminine, or neuter categories.

Gerund   See **Verbal.**

Helping verb   See **Auxiliary verb.**

Imperative mood   See **Mood.**

Indefinite pronoun   A pronoun not pointing out a particular person or thing. Some of the most common are *any, anybody, anyone, each, everybody, everyone, neither, one,* and *some.*

Independent (main) clause   A group of words that contains a subject and a predicate and that grammatically can stand alone as a sentence.

Indicative mood   See **Mood.**

Indirect object   A word that indirectly receives the action of the verb. See p. 20.

> The actress wrote the *soldier* a letter.

Infinitive   See **Verbal.**

Inflection   A change in the form of a word to indicate its grammatical function. Nouns, adjectives, and pronouns are inflected by declension; verbs, by conjugation. Some inflections occur when *-s* or *-es* is added to nouns or verbs or when *'s* is added to nouns.

Intensifier   A modifier (such as *very*) used to lend emphasis. Use sparingly.

Intensive pronoun   A pronoun ending in *-self* and used for emphasis.
   The director *himself* will act the part of Hamlet.

Interjection   A word used to exclaim or to express an (usually strong) emotion. It has no grammatical connections within its sentence. Some common interjections are *oh, ah,* and *ouch.* See p. 14.

Interrogative pronoun   See **9i.**

Intransitive verb   See **Voice.**

Inversion   A change in normal word order, such as placing an adjective after the noun it modifies or placing the object of a verb at the beginning of a sentence.

Irregular verb   A verb that does not form its past tense and past participle by adding *-d* or *-ed* to its infinitive form—for example, *give, gave, given.*

Linking verb   A verb that does not express action but links the subject to another word that names or describes it. See pp. 7, 19, and **10c.** Common linking verbs are *be, become,* and *seem.*

Main clause   See **Independent clause.**

Mixed construction   A sentence with two or more parts that are not grammatically compatible.
   MIXED CONSTRUCTION
      By cutting welfare benefits will penalize many poor families.
   REVISED
      Cutting welfare benefits will penalize many poor families.

Modifier   A word (or word group) that limits or describes another word. See pp. 98–101.

Mood   The mood (or mode) of a verb indicates whether an action is to be thought of as fact, command, wish, or condition contrary to fact. Modern English has three moods: the indicative, for ordinary statements and questions; the imperative, for commands and entreaty; and the subjunctive, for certain idiomatic expressions of wish, command, or condition contrary to fact.

INDICATIVE

> *Does* she *play* the guitar?
> She *does*.

IMPERATIVE

> *Stay* with me.
> *Let* him stay.

The imperative is formed like plural present indicative, without *-s*.

SUBJUNCTIVE

> If I *were you*, I would go.
> I wish he *were* going with you.
> I move that the meeting *be* adjourned.
> It is necessary that he *stay* absolutely quiet.
> If this *be* true, no man ever loved.

The most common subjunctive forms are *were* and *be*. All others are formed like the present-tense plural form without *-s*.

Nominal   A term for a word or a word group that is used as a noun—for example, the *good*, the *bad*, the *ugly*.

Nominative case   See **Case**.

Nonrestrictive modifier   A modifier that is not essential to understanding. See **20e**.

Noun   A word that names and that has gender, number, and case. There are proper nouns, which name particular people, places or things *(Thomas Jefferson, Paris,* the *Colosseum);* common nouns, which name one or more of a group *(alligator, high school, politician);* collective nouns (see **7d** and **8c**); abstract nouns, which name ideas, feelings, beliefs, and so on *(religion, justice, dislike, enthusiasm);* concrete nouns, which name things perceived through the senses *(lemon, hatchet, worm).*

Noun clause   See **Dependent clause**.

Number   A term to describe forms that indicate whether a word is singular or plural.

Object of preposition   See **Preposition, 9b**, and p. 20.

Objective case   See **Case**.

Parallelism   Parallelism occurs when corresponding parts of a sentence are similar in structure, length, and thought.

FAULTY

> The staff was required to wear black shoes, red ties, and *shirts that were white*.

PARALLEL

The staff was required to wear black shoes, red ties, and white shirts.

Participle   See **Verbal.**

Parts of speech   See pp. 2–14.

Passive voice   See **Voice.**

Person   Three groups of forms of pronouns (with corresponding verb inflections) used to distinguish between the speaker (first person), the person spoken to (second person), and the person spoken about (third person).

Personal pronoun   A pronoun like *I, you, he, she, it, we, they, mine, yours, his, hers, its, ours, theirs.*

Phrase   A group of closely related words without both a subject and a predicate. There are subject phrases *(the new drill sergeant)*, verb phrases *(should have been)*, verbal phrases *(climbing high mountains)*, prepositional phrases *(of the novel)*, appositive phrases (my brother, *the black sheep of the family*), and so forth. See pp. 21–23.

Predicate   The verb in a clause (simple predicate) or the verb and its modifiers, complements, and objects (complete predicate). See pp. 17–18.

Predicate adjective   An adjective following a linking verb and describing the subject. See p. 19 and **10c.**

The rose is *artificial.*

Predicate nominative   A noun following a linking verb and naming the subject.

The flower is a *rose.*

Predication   See **Faulty predication.**

Preposition   A connective that joins a noun or a pronoun to the rest of a sentence. See pp. 12–13, 22.

Principal parts   The verb forms **present** *(smile, go)*, **past** *(smiled, went)*, and **past participle** *(smiled, gone)*. See pp. 38–40.

Pronominal adjective   An adjective that has the same form as a possessive pronoun (*my* book, *their* enthusiasm).

Pronoun   A word that stands for a noun. See **Demonstrative pronoun; Indefinite pronoun; Intensive pronoun; Interrogative pronoun; Personal pronoun; Reflexive pronoun; Relative pronoun.**

Reflexive pronoun   A pronoun ending in *-self* and indicating that the subject acts upon itself. See **Myself** (Glossary of Usage).

Relative pronoun   See **9i**.

Restrictive modifier   A modifier essential for a clear understanding of the element modified. See **Nonrestrictive modifier.**

Run-on sentence   See **Fused sentence.**

Sentence fragment   See **Fragment.**

Sentence modifier   A word or group of words that modifies the rest of the sentence (*for example, frankly, in fact, on the other hand,* and so forth).

Simple sentence   A sentence consisting of only one independent clause and no dependent clauses. See p. 26.

Split infinitive   Infinitive with an element interposed between *to* and the verb form (*to* **highly** *appreciate*). Avoid. See p. 104.

Subject   A word or group of words about which the sentence or clause makes a statement. See pp. 15–16.

Subjective case   See **Case.**

Subjunctive mood   See **Mood.**

Subordinate clause   See **Dependent clause.**

Subordinating conjunction   A conjunction that connects a subordinating clause to the rest of the sentence. Some common subordinating conjunctions are *after, although, as, as if, as long as, as soon as, because, before, if, in order that, since, so that, though, unless, until, when, where, whereas, while.* See p. 11.

Substantive   A noun or a sentence element that serves the function of a noun.

Superlative degree   See **10a** and **10b**.

Syntax   The grammatical ways in which words are put together to form phrases, clauses, and sentences.

Transitive verb   See **Voice.**

Verb   A word or group of words expressing action, being, or state of being. See pp. 7, 38–45.

>Automobiles *burn* gas.
>What *is* life?

Verb phrase   See **Phrase**.

Verbal   A word derived from a verb and used as a noun, an adjective, or
an adverb. A verbal may be a gerund, a participle, or an infinitive. See
pp. 22–23.

Voice   Transitive verbs have two forms to show whether their subjects
act on an object (active voice) or are acted upon (passive voice). See
pp. 47–48.

(*credits continued from page vi*)

Winston Churchill. Reprinted with permission of Curtis Brown Ltd. on behalf of The Estate of Sir Winston Churchill, and Charles Scribner's Sons, an imprint of Macmillan Publishing Company, from *My Early Life: A Roving Commission* by Winston Churchill. Copyright 1930 Charles Scribner's Sons; copyright renewed © 1958 Sir Winston Churchill.

Betty Edwards. From *Drawing on the Right Side of the Brain* by Betty Edwards. Copyright © by Betty Edwards, Jeremy P. Tarcher, Inc., Los Angeles. Reprinted by permission of St. Martin's Press and Souvenir Press Ltd.

Willard Gaylin. "What You See Is the Real You" by Willard Gaylin, *The New York Times,* October 7, 1977. Copyright © 1991 by The New York Times Company. Reprinted by permission.

Gerard J. Gormley. Extract from a letter reprinted by permission of Mr. Gormley.

Buck Henry. Excerpt from screenplay to *The Day of the Dolphin* by Buck Henry. Reprinted by permission.

David H. Hubel. Excerpt from "The Brain" by David H. Hubel, *Scientific American,* September 1979, p. 45. Reprinted by permission.

Mary Oliver. "The Black Walnut Tree" from *Twelve Moons* by Mary Oliver. Copyright © 1978 by Mary Oliver. First appeared in *The Ohio Review.* By permission of Little, Brown and Company.

Gail Sheehy. Excerpts from *Passages* by Gail Sheehy are reprinted by permission of Penguin USA.

Douglas L. Wilson. Copyright 1991, Douglas L. Wilson. Used by permission of the author. "What Jefferson and Lincoln Read" first appeared in *The Atlantic Monthly,* January 1991.

Index

485

Abbreviations   Used in Marking Papers

GENERAL EDITING MARKS

awk — awkward

⌒ — close up

— delete

× — obvious error

∧ — insert

— insert space

o — omission

¶ — paragraphing

∿ — transposed